Parkinson's Disease Management through ICT: The REMPARK Approach

RIVER PUBLISHERS SERIES IN BIOMEDICAL ENGINEERING

Series Editor

Dinesh Kant Kumar
RMIT
Australia

Indexing: All books published in this series are submitted to Thomson Reuters Book Citation Index (BkCI), CrossRef and to Google Scholar.

The "River Publishers Series in Biomedical Engineering" is a series of comprehensive academic and professional books which focus on the engineering and mathematics in medicine and biology. The series presents innovative experimental science and technological development in the biomedical field as well as clinical application of new developments.

Books published in the series include research monographs, edited volumes, handbooks and textbooks. The books provide professionals, researchers, educators, and advanced students in the field with an invaluable insight into the latest research and developments.

Topics covered in the series include, but are by no means restricted to the following:

- Biomedical engineering
- Biomedical physics and applied biophysics
- Bio-informatics
- Bio-metrics
- Bio-signals
- Medical Imaging

For a list of other books in this series, visit www.riverpublishers.com

Parkinson's Disease Management through ICT: The REMPARK Approach

Editors

Joan Cabestany

Universitat Politècnica de Catalunya UPC
Barcelona, Spain

Àngels Bayés

Centro Médico Teknon, Grupo Hospitalario Quirón, Parkinson Unit
Barcelona, Spain

LONDON AND NEW YORK

Published 2017 by River Publishers

River Publishers
Alsbjergvej 10, 9260 Gistrup, Denmark
www.riverpublishers.com

Distributed exclusively by Routledge

4 Park Square, Milton Park, Abingdon, Oxon OX14 4RN
605 Third Avenue, New York, NY 10158

First published in paperback 2024

Parkinson's Disease Management through ICT: The REMPARK Approach / by Joan Cabestany, Àngels Bayés.

Routledge is an imprint of the Taylor & Francis Group, an informa business

Publisher's Note
The publisher has gone to great lengths to ensure the quality of this reprint but points out that some imperfections in the original copies may be apparent.

While every effort is made to provide dependable information, the publisher, authors, and editors cannot be held responsible for any errors or omissions.

ISBN: 978-87-93519-46-6 (hbk)
ISBN: 978-87-7004-426-4 (pbk)
ISBN: 978-1-003-33903-8 (ebk)

DOI: 10.1201/9781003339038

Contents

Preface

Parkinson's Disease (PD) is a neurodegenerative disorder that manifests with motor and non-motor symptoms. PD treatment is symptomatic and tries to alleviate the associated symptoms through a correct adjustment of the medication. As the disease evolves with time and this evolution strongly depends on each patient, it could be very difficult to correctly manage the disease.

Currently available ITC technology (electronics, communication, computing...) correctly combined with wearables can be of great utility to obtain and process useful information for both clinicians and patients. In this way, patients can actively involve in their condition and clinicians can obtain complimentary information for helping purposes.

This book presents the work done, main results, and conclusions of the REMPARK project (2011–2015) funded by the European Union under contract FP7-ICT-2011-7-287677. REMPARK system was proposed and developed as a real personal health device for the remote and autonomous management of Parkinson's Disease, composed of different levels of interaction with the patient, clinician and carers; and integrating a set of interconnected sub-systems: sensor, auditory cueing, smartphone, and server.

The sensor subsystem, using embedded algorithmics, is able to detect the motor symptoms associated with PD in real-time. This information is sent through the smartphone to the REMPARK server and is used for an efficient management of the disease.?

Implementation of REMPARK will increase the independence and Quality of Life of patients; and improve their disease management, treatment and rehabilitation.

List of Contributors

Albert Sama *Universitat Politècnica de Catalunya – UPC, CETpD – Technical Research Centre for Dependency Care and Autonomous Living, Vilanova i la Geltrú (Barcelona), Spain*

Alejandro Rodriguez-Molinero *National University of Galway – NUIG, Galway, Ireland*

Ana Correira de Barros *Fraunhofer Portugal AICOS – Assistive Information and Communication Solutions, FhP-AICOS, Porto, Portugal*

Angels Bayés *Centro Médico Teknon – Grupo Hospitalario Quirón, Parkinson Unit, Barcelona, Spain*

Anna Prats *Centro Médico Teknon – Grupo Hospitalario Quirón, Parkinson Unit, Barcelona, Spain*

Berta Mestre *Centro Médico Teknon – Grupo Hospitalario Quirón, Parkinson Unit, Barcelona, Spain*

Carlos Pérez *Universitat Politècnica de Catalunya – UPC, CETpD – Technical Research Centre for Dependency Care and Autonomous Living, Vilanova i la Geltrú (Barcelona), Spain*

Daniel Rodriguez-Martin *Universitat Politècnica de Catalunya – UPC, CETpD – Technical Research Centre for Dependency Care and Autonomous Living, Vilanova i la Geltrú (Barcelona), Spain*

Hadas Lewy *Maccabi Healthcare Services, Tel-Aviv, Israel (actually with Holon Institute of Technology, Holon, Israel)*

J. Manuel Moreno *Universitat Politècnica de Catalunya – UPC, CETpD – Technical Research Centre for Dependency Care and Autonomous Living, Vilanova i la Geltrú (Barcelona), Spain*

Joan Cabestany *Universitat Politècnica de Catalunya – UPC, CETpD – Technical Research Centre for Dependency Care and Autonomous Living, Vilanova i la Geltrú (Barcelona), Spain*

João Cevada *Fraunhofer Portugal AICOS – Assistive Information and Communication Solutions, FhP-AICOS, Porto, Portugal*

Jordi Rovira *Telefónica I+D, Spain (actually with MySphera, Valencia, Spain)*

M. Cruz Crespo *Centro Médico Teknon – Grupo Hospitalario Quirón, Parkinson Unit, Barcelona, Spain*

Ricardo Graça *Fraunhofer Portugal AICOS – Assistive Information and Communication Solutions, FhP-AICOS, Porto, Portugal*

Roberta Annicchiarico *IRCCS Fondazione Santa Lucia (FSL), Rome, Italy*

Rui Castro *Fraunhofer Portugal AICOS – Assistive Information and Communication Solutions, FhP-AICOS, Porto, Portugal*

Sheila Alcaine *Centro Médico Teknon – Grupo Hospitalario Quirón, Parkinson Unit, Barcelona, Spain*

Timothy J Counihan *National University of Galway – NUIG, Galway, Ireland*

Vânia Guimarães *Fraunhofer Portugal AICOS – Assistive Information and Communication Solutions, FhP-AICOS, Porto, Portugal*

List of Figures

List of Tables

List of Abbreviations

μC	Microcontroller
μSD	Micro Secure Digital
A2DP	Advanced Audio Distribution Profile
AAL	Ambient Assisted Living
ACS	Auditory Cueing System
ADL	Activity of Daily Living
AI	Artificial Intelligence
AIMS	Abnormal Involuntary Movement Scale
BADL	Basic Activities of Daily Living (e.g. bathing, toileting, feeding, transferring, dressing)
BAN	Body Area Network
BPM	Beats per minute
CDS	continuous dopaminergic stimulation
CP	Clinical Protocol
CPU	Central Processing Unit
CRF	Case Report Form
DAO	Data Access Object
DBS	deep brain stimulation
DG	Design Guideline
DMA	Direct Memory Access
DMA	Disease Management Application
DMS	Disease Management System
DSM	Diagnostic and Statistical Manual
DSP	Digital Signal Processor
EC	Ethical Committee
EHR	Electronic Health Record
FES	Functional Electrical Stimulation
FFT	Fast Fourier Transform
FOG	Freezing Of Gait
FP7	Framework Programme 7
GP	General Practitioners

GUI	Graphical User Interface
H&Y	Hoehn and Yahr
HCI	Human-Computer Interaction
HER	Electronic Health Record
HTTPS	HyperText Transfer Protocol over SSL
I/O	Input/Output
IADL	Instrumental Activity of Daily Living
IC form	Informed Consent Form
ICT	Information and Communication Technologies
IMU	Inertial Measurement Unit
IQR	Interquartile Range
JSON	JavaScript Object Notation
LED	Light Emitting Diode
LSVT	Lee Silver voice treatment
MA	Medical Application
MC	Motor complication
MDT	Multidisciplinary treatment
MG	Mobile Gateway
MIPS	Mega-instructions per second
MMSE	Mini Mental State Examination
MS	Milli Seconds
URL	Uniform Resource Locator
MS	Motor Symptom
MVC	Model-View-Controller
SQL	Standardized Query Language
MWI	Medical Web Interface
NE	Non evaluated (state)
NMC	Non-motor complication
NMS	Non-motor Symptom
NMSQ	Non Motor Symptoms Questionnaire
OECD	Organisation for Economic Co-operation and Development
PD	Parkinson's Disease
PHS	Personal Health System
PHS	Personalized Health System
PT	Physical Therapy
PwP	People with PD
QoL	Quality of Life
QoLRH	quality of life related to health

QUEST	Quebec User Evaluation of Satisfaction with assistive Technology
RBD	REM sleep behaviour disorder
RBF	Radial Basis Function
RCS	REMPARK Central Server
RE	Rule Engine
REM	rapid eye movement
RFCOMM	Radio Frequency Communication
SP	Subcutaneous Pump
SPM	Steps per minute
SUS	System Usability Scale
TSN	Transaction Sequence Number
UI	User Interface
UPDRS	Unified Parkinson's Disease Rating Scale
UX	User eXperience

1

Parkinson's Disease Management: Trends and Challenges

Angels Bayés[1] and Timothy J Counihan[2]

[1]Centro Médico Teknon – Grupo Hospitalario Quirón, Parkinson Unit, Barcelona, Spain
[2]National University of Galway – NUIG, Galway, Ireland

1.1 Introduction

Parkinson's Disease (PD) is a common neurodegenerative disorder that manifests with motor symptoms (MS) and non-motor symptoms (NMS). These symptoms vary from one patient to another and throughout the course of the disease, affecting patients' quality of life (QoL) progressively. Advanced PD represents a public health problem, given that it leads to the reduction in the capacity for self-care and deterioration of the QoL of those affected and their caregivers. The economic cost, including lost productivity and informal care, is about 20 billion euros in the world today.

Parkinson's Disease treatment is symptomatic and aims to alleviate the symptoms associated with the disease, through the precise adjustment of medication. The most widely used drug, levodopa, is effective usually across the lifespan. However, the onset of motor complications (MCs) a few years after starting treatment (e.g., ON–OFF fluctuations and dyskinesias) detract from its value. Symptomatic control of these complications is difficult and must be often optimized because the obtained improvement after such adjustment is not stable over the long term.

Treatment is primarily addressed to reduce the time that the patient is in the OFF state (i.e., time without medication effects), while avoiding the appearance of MCs and NMS, such as hallucinations or impulse control disorder. Reducing OFF hours is, therefore, one of the main parameters used to evaluate the effectiveness of therapeutic interventions, both in medical practice and in clinical trials. Gathering accurate information about the patient's condition throughout the day is essential in order to determine the optimal treatment plan. In clinical practice, the only method available is based on diaries filled in by patients and their caregivers about the ON/OFF hours and dyskinesias. However, this method has certain limitations that make unreliable medium and long-term monitoring: motor difficulties, memory failures that hinder regular compliance and subjective evaluation. Therefore, solutions that can improve disease management are of great interest and occupy important part of current research.

Another important aspect of the symptomatic treatment of PD is the multidisciplinary treatment (MDT). The multiple impairments occurring in PD have diverse functional and psychosocial consequences. While the primary treatment is pharmacological, many symptoms do not respond to medication, such as on-period freezing and postural instability. Indeed, later stage disease may be dominated by such symptoms. In addition, there is growing evidence for the efficacy of rehabilitation therapies for specific symptoms, through the involvement of the multidisciplinary team. There is also emerging evidence for physiotherapy with external cueing for improving gait and balance; cognitive movement strategies; and strength and balance exercises. Intensive speech therapy [e.g., Lee Silverman Voice Treatment (LSVT)] has been shown to improve the loudness and intelligibility of speech in PD. Unfortunately, the MDT is only applied in a small number of PD patients for economical and logistic reasons.

1.2 Impact and Strategies of PD at Different Stages

Parkinson's Disease is one of the chronic disorders with the most impact on patients' lifestyle. Most patients survive many years after the first symptoms. The mean survival rate of patients with this disease (when diagnosed after age 50) is 26 years, not very different from the non-affected population.

This disease responds very well to treatment with levodopa and dopaminergic agonists during the first years (between 3 and 7 years). As the disease progresses, the patient encounters a limitation of the effect of medical treatment due to the appearance of motor and non-motor complications. These entail a progressive difficulty in carrying out activities of daily living and leading an independent life. During the first year, the doctor establishes the possible diagnosis and starts one treatment. In course of 2–4 years, there is a relative normality and the medication is generally effective. Between the years 5 and 9, the effectiveness of medication usually decreases and treatment may need to be modified. Problems with driving, finances, and work may appear

Table 1.1 Modified Hoehn and Yahr scale

Scale	Description
1.0	Unilateral involvement only
1.5	Unilateral and axial involvement
2.0	Bilateral involvement without impairment of balance
2.5	Mild bilateral disease with recovery on pull test
3.0	Mild to moderate bilateral disease; some postural instability; physically independent
4.0	Severe disability; still able to walk or stand unassisted
5.0	Wheelchair bound or bed ridden unless aided

at this time. During years 10–13, there is an increasing disability: 60–75% of patients present some intellectual deficit, worsening immobility, incontinence, and increased risk of falling.

We can distinguish five evolutionary stages of the disease, although patients may not go through all of them. These are the stages of Hoehn and Yahr (Table 1.1).

The main problems presented by patients in the different evolutionary phases and the strategies currently recommended are explained in the following subsections.

1.2.1 Patients in Early Stages

In stage 1 of PD, facial expression is generally normal and also the posture. Tremor of a limb is the most common initial manifestation. It is often quite annoying, and it is the symptom that draws the attention of both the doctor and the patient. Tremor rarely interferes with the activities of daily living (ADL), although it disturbs and distresses the patient. Patients sometimes report difficulties in performing activities such as buttoning, typing, or cutting food. In the careful exploration of these patients, other parkinsonian signs in a limb, such as bradykinesia or slow movement, and stiffness, which contribute to these fine motor difficulties, are detected in addition to tremor. Decreased arm swing or dragging of a leg when walking can also be observed. These symptoms, often present for several years, are better tolerated than tremor.

In Stage 2 of PD, the involvement is bilateral. There may be loss of facial expression with decreased blinking. Slight flexion of the body may be present and, in general, arm swing when walking is diminished, without altering balance. Patients slow down when performing ADLs, and they require more time to dress, clean themselves up, get up from a chair, or tie their shoes on their own.

Depressive symptoms are also frequent, and these are detected in between 30–50% of the cases. Medical treatment, which is administered according to the severity of the symptoms, often produces side effects.

In these initial stages, patients are advised to learn about the disease, learn to speak naturally about their problems, learn to share difficulties, and go to the doctor accompanied by someone. Standardized psychoeducational programs, such as the "Edupark" program [1], are of great help during this transition. From the diagnosis, it is recommended to initiate MDT, which includes physical exercise and cognitive stimulation. It is better for patients to continue doing things by themselves, even if it is slowly, without rushing and with enough time. It is advisable to adapt the setting in which patients have to perform their ADLs and to be physically and mentally as active as possible.

Family members should also be informed and should know how to convey their support. It is recommended to see a doctor if depressive symptoms or side effects occur with medications.

1.2.2 Moderately Affected Patients

People with PD in stages 3 and 4 already have a degree of moderate–severe disability, as they experience difficulty walking and with balance. They explain that their gait is shortened, and that sometimes they have difficulties to make turns while they walk, in the corners of the rooms, or to cross doorways. Balance problems can cause falls. Sometimes while walking, they develop freezing of gait (FOG), or difficulty to stand, either forward, propulsion, or backward, retropulsion. The feeling of fatigue is a very frequent symptom. They have the feeling of needing a lot more effort to perform certain tasks, and often notice pains in the cervical, lumbar, or shoulder region. Symptoms of autonomic dysfunction may also be present in the form of orthostatic hypotension, extreme sensations of heat or cold, sweating not related to physical activity, sometimes in the form of crisis, and urinary or sexual dysfunction.

Many patients, at stage 3 or 4, experience side effects to chronic dopaminergic medication. The most annoying side effect for patients is the ON–OFF phenomenon. This phenomenon is often disabling and causes fear and insecurity. During the ON phase, patients can enjoy good mobility and carry out activities outside the home, such as shopping or social activities. However, during the OFF phase the patient may be completely disabled, with difficulty walking, getting up from a chair, or manipulating objects with hands. The appearance of OFF phases limits the social activities of the patient, often preventing them from going out. In this state, patients may find themselves in really dangerous situations, such as if this phenomenon occurs when crossing a street.

Dyskinesia, or involuntary movements, are another important problem that many patients present with during stages 3 and 4. In general, they have a choreiform nature: creeping movements of the extremities, or masticatory movements of the lower jaw, protrusion of the tongue, oscillations as they walk, and head and neck movements. Dyskinesias are a secondary symptom of dopaminergic medication, which usually occurs during the levodopa peak dose. If they are mild, the family is more aware of these movements than the patients themselves, who usually associate it with free time of parkinsonian symptoms. When they are severe, they can become incapacitating as much as the symptoms themselves.

Non-motor symptoms may appear in the form of sleep disturbances, vivid dreams, and nocturnal vocalizations. Night-time vocalizations, reported by the bed partner, consist of loud cries during sleep often accompanied by agitation of arms and legs. It is called "Rapid eye movement (REM) behaviour disorder". These events can disrupt sleep. Other frequent behavioural disorders in these stages are visual hallucinations, delusional ideas and confusional states of the paranoid type. Visual hallucinations in general are not very threatening in PD. They often describe the vision of family members, animals, or shadows that become animated objects.

The strategies recommended in these phases are aimed at understanding the MC's and NMC's and know how to monitor them. This will allow the patient to adjust the activities in each period. In case of ON-OFF fluctuations, dyskinesias, clinical worsening or behavioural disorders appear, the patient may inform the neurologist who will consider the possibility of a drug adjustment. It is, therefore, important to learn to do the patient's diary. This information will be crucial to optimize pharmacological treatment.

Patients, in these phases, should continue to maintain an active life and perform MDT, such as physical exercise, occupational therapy, speech therapy, and cognitive stimulation, according to individual needs. It is also recommended to the patient to continue doing things by himself, as long as possible.

1.2.3 Severely Affected Patients

Patients with PD, stage 5, are severely affected. They are usually confined to a wheelchair or in bed and require great assistance to make transfers. They are totally dependent for the realization of ADLs and have a great limitation on a personal level. Difficulties in speech and voice are often accentuated: these patients are often difficult to understand due to their low volume and poor articulation of words. They may eventually develop contractures and present decubitus ulcers or recurrent urinary tract infections.

Since the emergence of effective therapies for the cure of this disease, not all patients reach a state of total dependence. However, they are experiencing a progressive reduction in ON time and an increase in dependency time. In the final stages of this disease, the presence of progressive dysphagia can cause recurrent aspiration pneumonia, which is a possible cause of death. Other conditions that may contribute to this outcome are infections of pressure ulcers or urinary tract.

Since a causal treatment of the disease is still not possible, the objective for an optimal treatment will be to obtain for the patient a good QoL and the maximum of independence possible. In the advanced stages, it is recommended to increase the hygiene, to take care of the mobilization, to adapt the feeding and above all to take care of communication. The LSVT method has demonstrated efficacy in the treatment of speech and speech disorders. However, in very severe situations, it is advisable to maintain communication, even if external technical support is necessary.

Possible conduct disorders should be addressed, while enhancing the hobbies and pleasures that can still take place, such as listening to music, reading, or watching movies.

Caregivers should make them feel their support, while they should seek a replacement that allows them to have their own space and thus, avoid the burden of care and better adaptation when the patient passes away.

1.3 QoL in PD

Quality of life means well-being or satisfaction with aspects of life that are important to the person according to social standards and personal judgments. Because of this latter characteristic (i.e., personal judgment), the QoL is understood differently by each person and, therefore, it is difficult to define. The World Health Organization (WHO) defines it as: "an individual perception of the position in a person's life, in the context of the culture and value system in which he lives, in relation to his goals, expectations, standards and concerns".

Quality of Life, as related to Health (QoLRH), is the self-perception and assessment of the impact that the disease has on a patient's life and what its consequences are [2]. This assessment is extremely important because, when it is not possible to cure, maintaining the QoL of the patient is a priority of medical care. Among the different components of QoLRH, main attention must be paid on:

- physical aspects, which are related to symptoms and are functional (i.e., ability to perform activities);
- mental aspects in relation to mood and cognition;
- social aspects such as family role or social relations; and
- economic aspects.

In a recent study by Winter et al. [3], a baseline and 3, 6 and 12 months' assessments were performed on 145 Parkinson's patients. The average annual cost was calculated at 20.095 € per patient. The direct costs involved an expenditure of 13.185 € on medication, 3.526 € on hospital care, and 3.789 € on residences. The indirect costs accounted for 34.5% of the total costs (6.937 €). The costs of home care for the family accounted for 20% of direct costs. Factors associated with a higher total cost were fluctuations, dyskinesias, and younger age.

To assess QoLRH in PD, 56 studies were reviewed [4]. The three most important factors determining QoLRH in PD were depression, stage of the disease, and the time that has elapsed since the onset of the disease.

In another study by Sławek et al. [5], performed with 100 patients, the most important predictor for poor QoL was depression, followed by motor complications. Motor complications, especially nocturnal akinesia and dyskinesia, significantly decrease the QoL of Parkinson's patients [6]. Not only can dyskinesias affect QoL, but they can also increase health costs in patients with PD. This should be taken into account when planning treatment [7].

Despite the high impact of motor symptoms in Parkinson's, non-motor symptoms seem to influence patients' QoL even more [8]. Non-motor symptoms tend to accumulate. The average was 10 symptoms per patient in the populations studied and symptoms tend to intensify over time. Depression, anxiety, fatigue, sleep disorders, pain, orthostatic hypotension, and profuse sweating are some of those that have shown an individual relationship with loss of QoL. In fact, any symptom that, due to its intensity, is installed as a central problem in the life of the patient has a direct and important impact on his or her QoL. For example: difficulty swallowing, persistent constipation, urinary urgency and night-time urination, delusions and hallucinations, memory problems, or a sense of choking when breathing. At the global level, the main factors influencing the poor QoL of those affected by PD are (in order):

1. Depression
2. Overall disease severity (Hoehn and Yahr stage)
3. Dyskinesia

4. ON–OFF fluctuations
5. Age
6. Insomnia
7. Tremor
8. Cognitive dysfunction

Another element that must be taken into account is the QoL in caregivers of patients with PD [9]. 40% of caregivers indicate that their health suffers from caring. Nearly half have increased depression, two-thirds report that their social life has suffered. The caregiver becomes burned out more (burden of care) if the patient has more disability, or affective problems, mental confusion, or falls. There is a correlation between those caregivers that are most affected and the degree of a patient's depression. The conclusion is that more attention should be given to caregivers' care, particularly in advanced stages and/or with psychiatric and fall complications. These findings demonstrate that the QoL of both the patient and the caregiver depends, to a great extent, on the inclusion of the burden of care as one of the problems associated with PD [10].

1.4 State of Art of Current Trends in PD Management

The current treatment of PD is symptomatic and is applied through pharmacological and/or surgical treatment, associated with MDT.

The pharmacological treatment of PD aims to alleviate the symptoms associated with the disease, through the precise adjustment of medication. During the first few years of treatment, dopaminergic drugs are usually very effective. When the ON–OFF phenomena are already present, this objective is reached essentially by reducing the time during which the patient is in the OFF state, while avoiding the appearance of MCs and NMS. The reduction of OFF time is, therefore, one of the main parameters used to evaluate the effectiveness of therapeutic interventions, both in medical practice and in clinical trials. To determine the optimal treatment plan, gathering accurate information about a patient's condition throughout the day is essential. In clinical practice, the method currently available is based on diaries filled in by patients and their caregivers, recording hours of ON–OFF and the presence of dyskinesia. However, this method has limitations that make it unreliable, such as motor difficulties, failures in memory and in compliance, and subjective evaluation. It is necessary to know precisely the effect of drugs on reducing OFF hours and increasing the ON hours in Parkinson's patients. Reliable tools are, therefore, needed for detecting the motor condition of patients.

In patients with advanced PD, in which it is very difficult to control MCs and NMS through pharmacological adjustment, interventional therapy strategies are increasingly applied. These treatment strategies are aimed at obtaining continuous dopaminergic stimulation (CDS), either by using an infusion pump to deliver medication or by deep brain stimulation (DBS). However, these techniques are expensive, and often difficult to manage by the patient. Well-designed clinical studies on these interventional therapeutic approaches provided evidence for the efficacy of DBS and CDS in advanced PD and opened new perspectives for their use in earlier disease stages also.

On the other hand, there is growing scientific evidence of the benefit of the application of MDT, such as physiotherapy, speech therapy (e.g., LSVT), occupational therapy, cognitive stimulation, and psychoeducation in the treatment of Parkinson's Disease. Intensive and multidisciplinary rehabilitation slows the progression of motor decay, and slows the need to increase treatment with levodopa, which is postulated to have a neuroprotective effect [11]. Therefore, the application of MDT from the moment of diagnosis seems of great interest. There are several studies of multidisciplinary care in PD comparing outcomes before and after the intervention. Outpatient multidisciplinary care programs have reported short-term improvements in UPDRS (Unified Parkinson Disease Rating Scale) motor score, gait speed and stride length, speech, depression and health-related QoLRH. Long-term improvements in motor function have also been reported, and the authors comment that a close collaboration among members of the multidisciplinary team was essential to obtain the best results.

Potential limitations to the implementation of effective MDT are: distance, insufficient expertise among health professionals, poor interdisciplinary collaboration, poor communication, and lack of financial support for a multidisciplinary team approach. Regular face-to-face team meetings are important for effective functioning of the team. These meetings allow sharing of pertinent information and ensure the team is working towards shared goals for any given patient. The meetings can be a forum and stimulus for staff education, driving up quality of care. This type of coordinated multidisciplinary approach is sometimes referred to as interdisciplinary.

Most hospitals in Europe do not have a multidisciplinary service for the care of people with PD. These types of therapies are expensive and in addition, their application requires patients to frequently go from one place to another. This entails a number of limitations, both economic and logistical, for those affected with PD before having access to these therapies.

1.5 Needs and Challenges for Optimal PD Management

Current management of advanced PD is complicated and problems arising from poor QoL affect many patients. In 2001, *the Committee on Quality of Health Care in America Institute of Medicine* provided an objective analysis on healthcare. The report listed six aims, proposing that health care should be: safe, effective, patient-centred, timely, efficient, and equitable. However, current care for PD in the US, Europe, and likely the majority of the world, frequently does not meet these six aims [12]. PD care is often not safe. Individuals with PD who are hospitalized are often subjected to delayed treatment, contraindicated medications, prolonged immobility, lengthy stays, and high mortality [13, 14]. There are some comprehensive and distributed PD care models that are quite effective, but only few patients receive such care. Many PD-related hospitalizations are likely preventable. The patient-centred care that is timely has been rarely studied. Despite the limited evidence, focus groups and surveys suggest that individuals with PD want more personalized information from multiple disciplines that is delivered remotely in a timely manner [15]. PD care is very inefficient. Patients and their caregivers spend hours travelling and waiting in the clinic for routine follow-up appointments or for the application of complementary therapies.

Finally, and perhaps what may be most concerning, there exists inequity of current PD care. A primary determinant of the care that will be received is where you live. In the US, 42% of individuals with PD older than 65 and up to 100% of individuals in some rural areas do not see a neurologist soon after diagnosis [16]. In Europe, the first right expressed in the European Parkinson's Disease Association Charter is care from a physician with a special interest in PD. However, 44% of Europeans do not see a PD specialist in the first 2 years after diagnosis. Beyond neurological care, access to specialist nurses, occupational therapists, and counsellors is often more limited [17]. In less wealthy countries, the situation is even worse. China only has approximately 50 movement disorder specialists to care for more than 2 million individuals with PD and Bolivia only has 15. A door-to-door epidemiology study found that none of the individuals identified with PD had ever seen a physician, much less received treatment.

We can make the treatment safer, effective, patient-centred, efficient, and equitable only with the application of two conditions: that the treatment is applied mostly in the patient's home and with the use of tools, based on new technologies: sensors, communication platforms, and smartphones. This will overcome economic barriers and physical distance.

The simple fact of detecting accurately and reliably the clinical condition of the patient can mean a significant advance in the QoL of the patient, as this will affect a much more accurate adjustment of medication. In addition, with the help of adequate platforms, many more patients, as well as their caregivers, will receive more specialized medical care, complementary therapies, and psychoeducation as often as necessary, regardless of where they live.

In addition, reliable detection of the motor status of PD patients throughout the day can drastically change the value of drug clinical trials. Finally, the careful selection of patients amenable to the semi-invasive therapy options becomes more and more important and should be timely. An interdisciplinary setting is required to account for optimal patient information and awareness, selection of best individual treatment modality, training of relatives and caregivers, management of complications, and follow-up care.

The application of this type of tool is also of great interest in this section [18].

From a clinical point of view, the development of new technologies in the management of Parkinson's Disease must be validated so that the improvement of the QoL related to health is the main objective. Symptom-monitoring tools should be based on these premises: to provide a valid and accurate parameter of a clinically relevant characteristic of the disease; to find evidence that the parameter has an ecologically relevant effect on the specific clinical application; that a target interval can be defined in which the parameter reflects the appropriate treatment response; and finally, that the implementation is simple to allow repetitive use [19].

Remote monitoring from devices, such as wearable sensors, smartphones, platforms, disease management applications, smart beds, wall-mounted cameras, smart glasses, and even utensils, can monitor a patient's symptoms and function objectively in their environment, facilitating the delivery of highly personalized care.

Another aspect to improve PD care is that the most of it must be delivered at home. Current care models frequently require older individuals with impaired mobility, cognition, and driving ability to be driven by overburdened caregivers to large, complex urban medical centres. Moving care to the patient's home would make PD care more patient-centred. Demographic factors, including aging populations, and social factors, such as the splintering of the extended family, will increase the need for home-based care. Technological advances, especially the ability to assess and deliver care remotely, will enable the transition of care back to the home. However, despite its promise, this next generation of home-based care will have to overcome barriers, including outdated insurance models and a technological divide. Once these barriers are addressed, home-based care will increase access to high-quality care for the growing number of individuals with PD.

Emerging care models will combine remote monitoring, self-monitoring, and multidisciplinary care to enable the provision of patient-centred care at home and decrease the need for in-clinic assessments.

The demand for in-home care is likely to grow as a result of demographic, economic, social, and technological factors. Both the absolute number and proportion of older individuals with PD will increase.

1.6 Conclusion

A system for PD management will be necessary in the near future. It must be able to reliably assess the symptoms, facilitate patient disease management, and give them independence and the best QoL. At the same time, the tools must help the patient to stay physically and mentally active as much as possible. Finally, they must provide the neurologist with disease management tools to optimize the treatment.

References

[1] Smith, M., and Simons, G. (eds). (2006). *Patient Education for People with Parkinson's Disease and Their Carers: A Manual.* Hoboken, NJ: Wiley.

[2] Martínez-Martín, P. (1998). *Calidad de Vida relacionada con la Salud En La Enfermedad de Parkinson.* Belgrade: Ars Médica

[3] Winter, Y, Balzer-Geldsetzer, M., Spottke, A., Reese, J. P., Baum, E., Klotsche, J., et al. (2010). Longitudinal study of the socioeconomic burden of Parkinson's disease in Germany. *Eur. J. Neurol.* 17, 1156–1163.

[4] Forjaz, M. J., Frades-Payo, B., Martínez-Martín, P. (2009). The current state of the art concerning quality of life in Parkinson's disease: II. Determining and associated factors. *Rev Neurol.* 16–31, 49, 655–660.

[5] Sławek, J., Derejko, M., and Lass, P. (2005) Factors affecting the quality of life of patients with idiopathic Parkinson's disease: a cross-sectional study in an outpatient clinic attendees. *Parkinsonism Relat. Disord.* 11, 465–468.

[6] Chapuis, S., Ouchchane, L., Metz, O., Gerbaud, L., and Durif, F. (2005). Impact of the motor complications of Parkinson's disease on the quality of life. *Mov. Disord.* 20, 224–230.

[7] Pechevis, M., Clarke, C. E., Vieregge, P., Khoshnood, B., Deschaseaux-Voinet, C., Berdeaux, G., et al.; Trial Study Group. Effects of dyskinesias in Parkinson's disease on quality of life and health-related costs: a prospective European study. *Eur. J. Neurol.* 12, 956–963.

[8] Martínez-Martín, P., Schapira, A. H., Stocchi, F., Sethi, K., Odin, P., MacPhee, G., et al. (2007). Prevalence of nonmotor symptoms in Parkinson's disease in an international setting; study using Non-Motor Symptoms Questionnaire in 545 patients. *Mov. Disord.* 22, 1623–1629.

[9] Schrag, A., Hovris, A, Morley, D., Quinn, N., Jahanshahi, M. (2006). Caregiver-burden in Parkinson's disease is closely associated with psychiatric symptoms, falls, and disability. *Parkinsonism Relat. Disord.* 12, 35–41.

[10] Schrag, A., Hovris, A., Morley, D., Quinn, N., and Jahanshahi, M. (2006). Caregiver-burden in Parkinson's disease is closely associated with psychiatric symptoms, falls, and disability. *Parkinsonism Relat. Disord.* 12, 35–41.

[11] Frazzitta, G. et al. (2015). Intensive rehabilitation treatment in early Parkinson's disease: a randomized pilot study with a 2-year follow-up. *Neurorehabil. Neural. Repair* 29, 123–131.

[12] Institute of Medicine (US) and Committee on Quality of Health Care in America. (2001). *Crossing the Quality Chasm: A New Health System for the 21st Century.* Washington, DC: National Academy Press.

[13] Gerlach, O. H., Winogrodzka, A., and Weber, W. E. (2011). Clinical problems in the hospitalized Parkinson's disease patient: systematic review. Mov Disord. 26, 197–208.

[14] Aminoff, M. J., Christine, C. W., Friedman, J. H., et al. (2011). Management of the hospitalized patient with Parkinson's disease: current state of the field and need for guidelines. *Parkinsonism Relat. Disord.* 17, 139–145.

[15] van der Eijk, M., Faber, M. J., Post, B., et al. (2015). Capturing patients' experiences to change Parkinson's disease care delivery: a multicenter study. *J. Neurol.* 262, 2528–2538.

[16] Willis, A. W., Schootman, M., Evanoff, B. A., Perlmutter, J. S., and Racette, B. A. (2011). Neurologist care in Parkinson disease: a utilization, outcomes, and survival study. *Neurology* 77, 851–857.

[17] Stocchi, F., and Bloem, B. R. (2013). Move for change part II: a European survey evaluating the impact of the EPDA Charter for people with Parkinson's disease. *Eur. J. Neurol.* 20, 461–472.

[18] Dorsey, E. R., Vlaanderen, F. P., Engelen, L., Kieburtz, K., Zhu, W., Biglan, K. M. (2016). Moving Parkinson care to the home. *Mov. Disord.* 31, 1258–1262.

[19] Maetzler, W., Klucken, J., and Horne, M. (2016). A clinical view on the development of technology-based tools in managing Parkinson's Disease. *Mov. Disord.* 31, 1263–1271.

2

Objective Measurement of Symptoms in PD

Timothy J Counihan[1] **and Angels Bayés**[2]

[1]National University of Galway – NUIG, Galway, Ireland
[2]Centro Médico Teknon – Grupo Hospitalario Quirón, Parkinson Unit, Barcelona, Spain

2.1 Advancing Parkinson's Disease: Motor and Non-Motor Fluctuations

The triad of rest tremor, slowness of movement (bradykinesia), and limb rigidity constitute the clinical hallmarks of idiopathic Parkinson's Disease (PD). However, the majority of patients will develop an increasing number of more complex symptoms over time. These symptoms include variability in the patient's mobility, so-called motor fluctuations, as well as variability in other non-motor symptoms (NMSs). Much of this variability is accounted for by a change in the patient's response to dopaminergic medication over time, such that the duration of benefit for a given dose of dopaminergic medication shortens. The following section will elaborate further on these motor and non-motor fluctuations and how they affect patients' ability to function. It is important to note, however, that not all motor fluctuations are the result of changing responses to dopaminergic medication.There is increasing recognition that several progressive motor symptoms such as postural instability and freezing of gait (FOG) are the result of non-dopaminergic dysfunction and, therefore, unlikely to respond to standard pharmacologic modifications. The clinical picture of advancing disease is further complicated by the emergence of other symptoms including cognitive dysfunction, sleep disturbances, and blood pressure instability, all of which combine to contribute to the increasing risk of falls in patients with PD. These emergent symptoms probably are a reflection of the underlying pathogenesis in PD, which is thought to involve spread of neuronal degeneration not only to dopaminergic neurons but also autonomic and cholinergic pathways. The result of advancing PD, therefore, is a tendency for patients to develop motor and non-motor fluctuations, postural and autonomic instability, cognitive impairment, and increasing falls risk. As we shall see in the following section, the task for the clinician is to accurately document the emergence of these symptoms, so that appropriate treatment strategies may be initiated to restore patients' quality of life (QoL).

2.1.1 Motor Fluctuations

Over 70% of patients with PD will develop increasing variability in their motor response to dopaminergic treatment after 5 years [1]. These "motor fluctuations" typically appear insidiously and are often initially unnoticed by the patient. The first evidence of an emerging fluctuating medication response is the "wearing-off" phenomenon, where the patient reports slowing up or increasing stiffness as the time approaches for their next dose of medication. This commonly is apparent overnight as the patient develops difficulty turning over and getting comfortable in bed (nocturnal akinesia). Patients report that they feel the first morning dose of medication "kick in" around half an hour after taking it ("morning benefit"). Over time, patients may notice end of dose "wearing-off". This may be ultimately quite dramatic to the point where they "freeze" and become effectively immobile until such time as their next dose "kicks in". Furthermore, the time needed for the subsequent dose to start working may be strongly influenced by several factors including dietary intake of protein, which may result in a delayed time before the patient turns ON. Motor fluctuations can become so severe as to result in the ON–OFF phenomenon, where the patient may cycle rapidly between being ON (mobile) and OFF (immobile or frozen). As mentioned above, these fluctuations are the result of alterations in dopamine receptor sensitivity to exogenous dopamine. The increasingly attenuated motor response to dopaminergic medication is further complicated by an exaggerated sensitivity of the motor response, resulting in excessive involuntary movements (dyskinesia), typically during the time of peak plasma levels of dopamine. Less commonly, painful twisting (dystonic) movements occur as dopamine levels rise or fall (biphasic dyskinesia). The OFF state may be so severe in patients as to cause painful "off dystonia" or FOG. It will be apparent, therefore, that there is a wide repertoire of variability in a given patient's motor state, ranging from complete immobility to flinging choreic movements, and that these fluctuations may occur cyclically many times throughout the day. Furthermore, the occurrence of motor fluctuations is influenced by many factors, including the dose and formulation of dopaminergic drug, diet, anxiety, and use of concurrent medications. It is essential that the treating clinician documents an accurate account of these motor fluctuations, so as to guide further treatment to minimize their disruption to the patient (Table 2.1).

Table 2.1 Common motor and non-motor fluctuations in PD

Motor	Non-motor
Wearing-off	Sensory
• Morning benefit • Nocturnal akinesia • ON/OFF	• Paresthesias • Hyposmia
Freezing of gait (FOG)	Autonomic
• Start hesitation • Festination	• Sweating • Constipation • Orthostatic Hypotension
Dyskinesia	Neuropsychiatric
• Peak dose • Biphasic	• Anxiety • Depression • Psychosis
OFF dystonia	Sleep
	• Rapid eye movement (REM) sleep behavior disorder • Excess daytime sleepiness

2.1.2 Non-Motor Fluctuations

It has been increasingly recognized that while much of the day to day disability in PD relates to motor dysfunction, many patients experience a variety of NMSs, including paresthesias, fluctuating mood, and anxiety, autonomic disturbances such as constipation, postural hypotension, sleep disturbances, as well as cognitive slowing (bradyphrenia) [2, 3]. As alluded to above, many of these symptoms reflect the underlying degenerative process, which affects several neurotransmitter systems. However, it is also apparent that many of these NMSs occur as a result of dopaminergic dysregulation, and are, therefore, fluctuant in the same way as motor symptom fluctuation. NMSs typically affect one or more of four domains: neuropsychiatric, autonomic, sensory, or sleep (Table 2.1). Many NMSs are now recognized to precede the development of motor symptoms in PD, and hence, can be seen as intrinsic to the disease process. Other NMSs are likely to be due to medication effects. The importance of recognizing fluctuations in non-motor dopaminergic symptoms (such as OFF-related paresthesias or anxiety) is that their management will be entirely different to similar symptoms that may not be dopaminergic in origin. For instance, many patients will develop acute onset anxiety symptoms as the first symptom of wearing-off of their dopaminergic medication. Treatment of this symptom will involve modifications in their l-dopa intake rather than prescribing anxiolytics. It will be again apparent that it is essential for the clinician to accurately document these symptoms so as make appropriate alterations in therapy may be made. Teasing out non-motor symptomatology can be challenging even for the experienced clinician. Adding to the complexity of this assessment is the fact that cognitive dysfunction develops in the majority of PD patients over time, so that after 15 years, 80% of patients experience significant cognitive impairment [4]. Moreover, chronic sleep disturbances due to nocturnal akinesia, rapid eye movement (REM) sleep behavior disorder and psychosis frequently result in excessive daytime somnolence. Each of these factors contributes to the difficulty that many patients have in recognizing fluctuations in MS and NMS.

2.2 Challenges in Documenting "Real-World" Fluctuating PD Symptoms

The complexity of evolving motor and non-motor fluctuations in PD poses a significant challenge to the treating clinician. It will be apparent from the previous discussion that accurate identification of true dopaminergic motor and non-motor fluctuations is paramount to the future management strategy for each patient. In the context of a busy hospital or clinic-based environment with its restrictions on the face-to-face patient time, as well as patient anxiety and fatigue, all contribute to the problem of accurate documentation of a patient's clinical status. The clinician has a number of well-worn skills at their disposal so as to become informed of how a patient is coping. The traditional history taking and detailed examination techniques provide invaluable information not only about the patient's motor function, but importantly about the patient's understanding of the different motor states. The latter is a critical aspect of the clinical encounter. Is the patient able to distinguish tremor (a manifestation of the OFF state) from dyskinesia (an ON phenomenon)? What is the patient's understanding of the terms OFF and ON? While the clinician examination will usually help in determining the patient's motor status, it is only a snapshot of a patient's daily life, and even at that, often not a particularly reliable one, as a visit to the doctor's office is not representative of a patient's daily activity, and the added burden of long travel to the clinic, a degree of patient anxiety about the encounter, as well as forgotten tablets, may all contribute to a distorted account of a patient's "real-life" clinical status.

A number of additional tools are available to help the clinician acquire the necessary data to inform therapeutic decision-making. The availability of an informed family member provides a useful narrative regarding the patient's home life, although too often the patient may live alone or a spouse is unavailable or unable to provide helpful information. A variety of clinical rating scales and questionnaires have been developed which provide a degree of objectivity to the patient's assessment, though all have their own limitations. Although the Unified Parkinson's Disease Rating Scale (UPDRS) remains the gold standard clinical rating scale especially in a research setting, it suffers from subjectivity and variable clinician competence in using it. For many years patient diaries have provided the backbone of "real-world" interrogation of patient motor states. Diaries allow for frequent recording of a patient's motor state (OFF, ON, ON with dyskinesia; and asleep) as determined by the patient or caregiver. However, patient reported diaries are known to lack consistency and can be associated with variable adherence. In addition, patients may have physical difficulty filling in the diary due to micrographia and severe akinesia. Similarly, many patients have difficulty distinguishing tremor from dyskinesia, or may erroneously record the state they are in at the moment of documentation, rather than their predominant motor state over the preceding hour or two. What is certainly clear in relation to the accurate assessment of motor fluctuations is that the patient's motor state as determined in the clinic is rarely representative of their quotidian state; indeed, it is not uncommon for the patient to cycle from extreme akinesia to severe dyskinesias during the course of the clinic encounter. The situation regarding accurate detection of non-motor fluctuations is even more difficult, and relies mostly on obtaining a careful history from the patient and their family regarding any possible periodicity of such symptoms (Table 2.1). The NMS questionnaire [5, 6] is a 30-questions assessment tool which can be useful in highlighting symptoms not previously recognized as potentially related to PD. Such questionnaires may lack the ability to identify non-motor fluctuations as part of a medication effect and are impractical for clinical routine use.

2.3 Emerging Technologies to Monitor Symptom Fluctuations

In an attempt to improve the ability to detect and monitor the occurrence of motor fluctuations, a variety of new technologies have emerged which is beginning to transform the management of Parkinson's Disease. Two technologies have recently been combined which promise to radically change the daily management of PD: so-called "wearables" and machine-learning based techniques [7]. Suffice it to say here that these technologies can provide objective, high frequency, sensitive, and continuous data on motor and non-motor phenomena in PD [7]. In particular, the development of wearable inertial sensors has introduced a level of objectivity in the recording of patients' motor function in daily life, so-called *free-living* monitoring [8]. The explosion of interest in the quantitative assessment and management of PD using technology-based tools has been usefully summarized in a series of reviews in the journal *Movement Disorders* [9]. Several large-scale studies on wearable technologies have reported preliminary results, along with over a dozen smaller series using a variety of machine learning algorithms for wearable sensor-based data acquisition in PD (summarized by Kubota et al. [7]). The obvious appeal of wearable technology (WT) is that it allows for community-based data acquisition over a continuous time period. In addition, the collected data is free from observer bias and the so-called attentional compensation (Hawthorne effect). The newer devices are relatively low cost with user-friendly technology. Continuous monitoring allows for the accumulation of large amounts of data, which may be analyzed at both a macro- or micro-level [10]. At a micro-level, data is available on motor state, frequency and severity of motor fluctuations, gait as well as response to medications. In the case of patients with PD, data can be acquired on the number and intensity of multiple activities, including the frequency and amplitude of movements throughout the day and night, the frequency and duration of tremor, dyskinesia, as well as impairments of gait including FOG and balance impairment [11]. There is also the possibility that detailed interrogation of data obtained at regular intervals may disclose important subclinical physiological changes that might predict a subsequent clinical change in motor state, allowing potentially for preemptive therapeutic intervention. This appears to be especially relevant in the detection of FOG [12, 13]. A big challenge, however, for any emerging sensor technology aiming to supersede existing clinical motor assessments (such as the UPDRS) is the requirement for clinimetric algorithm validation. Testing of such objective algorithms must include a test/re-test protocol to ensure reliability.

At a macro-level, general levels of activity, inactivity or sleep are available. The information obtained from WT is available both to the clinician and the patient, and importantly it allows the patient to become actively involved in their condition and learn about their level of activity as well as therapeutic efficacy. An important aspect of a WT system is that it allows remote monitoring of symptoms with its obvious potential advantage for patients and health economics (REMPARK system, presented in this book, is an example).

It should be apparent that the ability to quantify clinical phenomena using WT is not simply because "*we can*", but that the ability to reliably capture such data is a major advance on the standard qualitative clinical examination. Espay et al. [11] point out that whatever is measured must be directed toward a therapeutic target. Despite these technological advances, however, their application in a clinical setting has been slower to evolve. As Del Din et al. [10] pointed out, there remains no gold standard against which remote monitoring WT can be compared. Moreover, simply measuring a litany of motor and non-motor phenomena should not necessarily lead to reflexly acting on such data with therapeutic intervention. What matters for one patient may cause little functional impairment for another. For instance, a common experience for the clinician is to observe a patient with severe, ballistic dyskinesias, but for whom the involuntary movements are non-disabling [14]. This dichotomy becomes even starker when one attempts to take into account various non-motor measurements (such as hyperhidrosis or anxiety), which may cause substantial disability for the patient, but remain less overt than motor recordings. A number of attempts have been made to record non-motor symptomatology using sensor technology, chiefly relating to sleep-related symptoms [15]. These sensors can provide valuable real-world insight into a variety of nocturnal sleep difficulties including RBD (REM sleep behavior disorder), nocturnal akinesia, and restless legs syndrome. Moreover, information on excessive daytime somnolence is also recordable on inertial devices. In recognition of this, several commentators have called for a "multidomain" integrated technology as the template for individualized, personalized therapeutic approach to care [11]. The REMPARK system is an attempt to provide an integrated system of personalized management of PD.

2.4 Challenges in the Use of Inertial Sensor Technology in Monitoring PD Symptoms

There exist some specific challenges in the use of technology for an objective monitoring of PD and the contribution to increase QoL of patients. A set of them are discussed below:

1. *Limitations of Motor Sensors*: Currently available commercial sensor technologies have been demonstrated to accurately measure a variety of motor symptoms in PD including tremor, bradykinesia, dyskinesia, and gait impairment [16]. However, much of the validation testing of such technology was derived from laboratory-based experimentation. Clinicians have raised a concern that this is not an accurate reflection of "real-world" situations. For example, sensor data fails to discriminate accurately whether bradykinesia is the result of fatigue, anxiety, or motor wearing-off [12]. Hence, the *context* in which sensor data is derived has a large influence on its interpretation. While the freedom to record

continuous data in an ambulatory home setting has obvious advantages over the laboratory, the trade-off is a substantial loss of experimental control over the data collection [7]. The REMPARK system and others have attempted to provide a degree of context to the sensor data by the provision of smartphone technology, which allows the patient or carer to interact with a web-based application to input subjective data, such as a non-motor questionnaire [13]. Another potential limitation of inertial sensor measurements is that their resolution may be affected by their anatomical location, and that while this may be mitigated to some extent by the application of multiple sensors, this adds to the complexity of the algorithms to interpret the data, as well as adding more complexity and discomfort for the patient [17]. Currently, most available sensors rely on single sensors applied to the most affected side in patients.

2. *NMS Monitoring*: The development of WT to monitor PD symptoms has primarily focused on motor aspects of the disease [11]. However, it has been apparent even using older established clinical scales of motor disability (UPDRS) that there is at best only a modest correlation between the degree of motor impairment and QoL. There is an emerging recognition that a substantial daily burden for patients derives from non-motor symptomatology [5], including symptoms such as anxiety, sweating, fatigue, sleep disturbances, etc. While some technologies are available to address this, such as sweat sensors, blood pressure and heart rate monitors, as well as the application of inertial sensors to monitor for sleep disturbances, clearly further refinement of this technology is needed.

3. *Clinimetrics and data validation*: There is a natural tendency to consider the development of "objective" sensor technology as ultimately replacing previously validated subjective rating scales such as the UPDRS. However, Espay et al. [11] have pointed out that the clinician integrates many sources of "data" to reach a subjective score or opinion, and that technological sensor measurements should complement clinical measures rather than act as surrogate markers.

4. *Disease Progression*: The advent of WT has opened the door to their use in clinical trials of emerging therapies as surrogate markers of efficacy. But investigators need to be cognizant that improved sensor "data" following institution of a therapy may not translate into improved QoL for patients. Moreover, use of WT longitudinally will need to take into account that PD is a progressive disease and that the parameters by which improvement must be measured will change over time. For instance, for many patients

with advanced disease, the priority of maintaining cognitive function may come at the expense of a desirable motor outcome, at which point motor sensor data becomes moot.

5. *Data Analysis*: Existing technology-based objective measures require a robust method of data analysis to ensure that the data collected is an accurate reflection of the clinical phenomena under scrutiny. The development of algorithms to appropriately interrogate the data requires close cooperation between clinicians and technical personnel. For instance, a sensor that detects "severe bradykinesia" at one measurement and "severe dyskinesia" at a second measurement within a two-minuteepoch is likely to be spurious, unless data from other sources can be factored in. In this case, the data suggesting "severe bradykinesia" is likely to be strengthened in its validity if such data is accompanied by a contemporaneous finding of "severe tremor". Development of reliable algorithms to interpret data requires input from clinicians so that appropriate weighting can be given to such data.

6. *User Engagement*: Despite enthusiasm from technologists and clinicians for the incorporation of WT into practice, current research suggests only modest patient and carer engagement [18]. Giving feedback to patients on the relevance and importance of wearing sensors seems to influence adherence.

2.5 Conclusion

There is little doubt that the rapid development of continuous WT with the ability to generate objective measurements relating to the evolving clinical state of PD patients has provided a tremendous opportunity to improve patient care, as well as learn more about the disease as it affects patients in their own environment. In the long term, these technologies will have to be sufficiently reliable and robust to allow for the remote personalized management of PD, including treatment delivery.

References

[1] Aquino,C. C., and Fox, S. H. (2015). The clinical spectrum of levodopa-induced complications. *Mov. Disord.* 30, 80–89.
[2] Lim, S.-Y., and Lang, A. E. (2010). The non-motor symptoms of Parkinson's disease: an overview. *Mov. Disord.* 25, S123–S130.

[3] Martinez-Fernandez, R., Schmitt, E., Martinez-Martin, P., and Krack P. (2016). The hidden sister of motor fluctuations in Parkinson's disease: a review on non-motor fluctuations. *Mov. Disord.* 31, 1080–1094.

[4] Hely, M. A., Reid, W. G., Adena, M. A., Halliday, G. M., and Morris, J. G. (2008). The Sydney Multicenter Study of Parkinson disease: the inevitability of dementia at twenty years. *Mov. Disord.* 23, 837–844.

[5] Chaudhuri, K. R., Martinez-Martin, P., Schapira, A. H. V., Stocchi, F., Sethi, K., and Odin, P. et al. (2006). An international multicentre pilot study of the first comprehensive self-completed non motor symptoms questionnaire for Parkinson's disease: The NMSQuest study. *Mov. Disord.* 21, 916–923.

[6] Romenets, S. R., Wolfson, C., Galatas, C., Pelletier, A., Altman, R., Wadup, L., Postuma, R. B., et al. (2012). Validation of the non-motor symptoms questionnaire (NMS-Quest). *Parkinsonism Relat. Disord.* 18, 54–58.

[7] Kubota, K. J., Chen, J. A., and Little, M. A. (2016). Machine learning for large scale wearable sensor data in Parkinsons disease; concepts, prospect's, pitfalls and futures. *Mov. Disord.* 31, 1314–1326.

[8] Lowe, S. A., and O'Laighin, G. (2014). Monitoring human health behavior in one's living environment: a technological review. *Med. Eng. Phys.* 36, 147–168.

[9] Sanchez-Ferro, A., and Maetzler, W. (2016). Advances in sensor and wearable technologies for Parkinson's disease. *Mov. Disord.* 31, 1257.

[10] Del Din, S., Godfrey, A., Mazza, C., Lord, S., and Rochester, L. (2016). Free-living monitoring of Parkinson's disease: lessons from the field. *Mov. Disord.* 31, 1293–1313.

[11] Espay, A. J., Bonato, P., Nahab, F. P., et al. (2016). Technology in Parkinson's disease: challenges and opportunities. *Mov. Disord.* 31, 1272–1282.

[12] Maetzler, W., Klucken, J., and Horne, M. (2016). A clinical view on the development of technology-based tools in managing Parkinson's disease. *Mov. Disord.* 31, 1263–1271.

[13] Ahlrichs C, Samà, A., Lawo, M., Cabestany, J., Rodríguez-Martín, D., Pérez-López, C. et al. (2016). Detecting freezing of gait with a tri-axial accelerometer in Parkinson's disease patients. *Med. Biol. Eng. Comput.* 54, 223–233.

[14] Hung, S. W., Adeli, G. M., Arenovitch, T., Fox, S. H., and Lang, A. E. (2010). Patient perception of dyskinesia in Parkinson Disease. *J. Neurol. Neurosurg. Psychiatry* 81, 1112–1115.

[15] Klingelhoefer, L., Rizos, A., Sauerbier, A., et al. (2016). Night-time sleep in Parkinson's disease; the potential use of Parkinsons KinetiGraph: a prospective comparative study. *Eur. J. Neurol.* 23, 1275–1288.

[16] van Uem, J. M. T., Maier, K. S., Hucker, S., et al. (2016). Twelve week sensor assessment in Parkinson's disease: impact on quality of life. *Mov. Disord.* 31, 1337–1338.

[17] Fisher, J. M., Hammerla, N. Y., Rochester, L., and Andras, P., and Walker, R. W. (2016). Body-worn sensors in Parkinson's disease: evaluating their acceptability to patients. *Telemed. J. E Health* 22, 63–69.

3

The REMPARK System

Joan Cabestany[1], J. Manuel Moreno[1] and Rui Castro[2]

[1]Universitat Politècnica de Catalunya – UPC, CETpD – Technical Research Centre for Dependency Care and Autonomous Living, Vilanova i la Geltrú (Barcelona), Spain
[2]Fraunhofer Portugal AICOS – Assistive Information and Communication Solutions, FhP-AICOS, Porto, Portugal

3.1 Introduction

As it has been presented and described in previous chapters, Parkinson's Disease is a progressive neurological condition. Unfortunately, there is no known cure for the disease and only treatments focused to the management and mitigation of the different symptoms are available to patients.

A well-designed and adapted technological approach (mainly based on WT and machine learning, as suggested in Chapter 2) can help both, the patients and their neurologists, for a better management and follow-up of the disease evolution. This will open the possibility to more effective therapies in order to improve patients' daily activities. From the beginning, with this specific aim, REMPARK project proposed four main objectives to be covered:

Objective 1: **Real-time motor status identification**.

To develop a minimum set of wearable inertial, electronic, intelligent sensors, to register in real-time the ambulatory movements of the patient and to clearly identify the relevant number of parameters, or types of motor disorders associated with Parkinson's Disease.

Objective 2: **Gait improvement system**.

To develop a self-calibrating gait guidance system to prevent the gait impairments and therefore minimize the appearance of FOG episodes and falls. The development will be mainly focused on auditory cueing, based on a system able to generate an auditory or acustic signal when necessary (Auditory Cueing System ACS), in order to help patients to improve their gait.

Objective 3: **User interface to get patient's feedback**.

To develop and implement a specific user interface on a smartphone to collect feedback directly from the patient. This feedback provided by the patient must have, at least, the following functionalities:

- Introduction of routine information such as the medications' intake time, quantity and quality of the sleep and other NMS.
- Answer to specific prompts to validate some situations detected by the REMPARK system. For instance, periodically completing medical tailored tests designed by the doctors (based on the UPDRS scale) in order to assess the evolution of the patient with time regarding the non-motor symptoms.

Objective 4: **REMPARK service for remote management of the disease**.

A central server service (REMPARK Server) organized to act as a repository of the processed ambulatory data of the patients, to ease the combination of these data with the Electronic Health Record (EHR) and to facilitate interaction with the neurologist (or medical service), the patient and/or the caregiver.

The present chapter introduces a deeper view of the presented approach, describing the details of the implemented and used system for the works carried out during the evolution of REMPARK project.

3.2 REMPARK System Overview

The developed REMPARK system has been worked-on and designed as a Personal Help System (PHS) that provides a closed-loop detection and treatment capabilities for an improved management of Parkinson's Disease (PD) patients.

From the beginning, the REMPARK architecture was conceived and organized in two different levels:

- **The immediate level (BAN – Body Area Network)**. A minimum set of wearable monitoring sensors should be able to identify in real time the motor status of the patient. The identification of the status is made autonomously using embedded algorithms in the sensor. The sensitivity and specificity of these sensors are better than 80%. A gait guidance system, based on an auditory cueing system ACS, was envisioned as part of this level to help the patients in real time during their daily activities. The immediate level is also including actuation capabilities that will be triggered by certain conditions detected by the sensors:
  - activation of the ACS;
  - automatic administration of a questionnaire;
  - possible remote control of a delivering medication pump.

  A possible inclusion of a remotely-controlled pump was done in REMPARK project as a laboratory exercise with no real tests in the trials with patients. This is a challenging work aiming to give a response to the need of a more accurate and adapted dose control for infusion pump systems. It must be considered as a step forward in the improvement of the QoL of patients with advanced PD.

  All the information from and to the immediate level are carried out by a smartphone which is used as a gateway.
- **The mid-term level** (based on a server). This is where the data provided by the sensors are partially analyzed, along with disease management done automatically or/and by neurologists. This part of the REMPARK system provides interconnection between analysis performed in the sensor and data provided by the neurologist, caregivers or the patient. An important tool at this stage is the Disease Management System (described in Chapter 8) that opens the possibility to a high level management of PD and to the establishment of interaction between concerned people.

These two levels and their relationship are represented in Figure 3.1. Below, Figure 3.2 shows a more detailed content of the described levels, together with the associated functionality. In these levels, two different groups of algorithms are running. The first group is located in the Immediate level (BAN), mainly in the sensor and smartphone, and it is responsible for the ON/OFF detection, the different symptoms identification and the implementation of the gait guidance system. A second group of algorithms is responsible of the implementation of the Rule Engine at the second level.

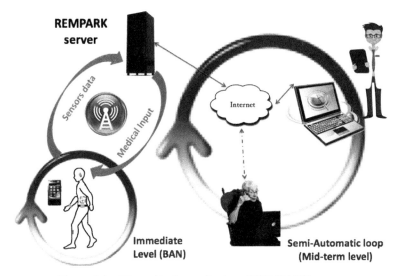

Figure 3.1 Hierarchical organization of REMPARK system.

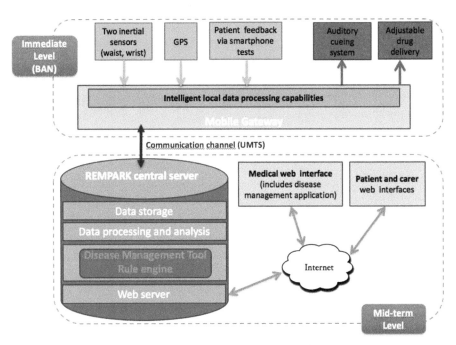

Figure 3.2 System architecture overview.

3.2.1 The Immediate Level

The Immediate Level corresponds to the Body Area Network (BAN) and acts in the short-term. It is composed of the sensors, the actuators and the smartphone, which also acts as GPS for providing context-aware information and as an interface for the PD patient, to record their direct feedback especially regarding the non-motor symptoms.

Communications among the elements of the BAN is done through Bluetooth V2.1, allowing enough autonomy for the patient. This level operates autonomously and it is a real operative closed loop. It is auto-adaptive by means of a constant evaluation of the actuator's effect, correcting its behavior in the short-term. The configuration of this level can be changed according to different patients' needs.

Considering the details indicated in the Figures 3.1 and 3.2, the main components or sub-systems having a direct interaction with the PD patient are the following:

- **Sensor module**: It is in charge, using appropriate technology, of the continuous monitoring of the patient's movement behaviour. It is able to detect and identify in real-time the different motor-related symptoms suffered by the patient.
- **Mobile gateway**: As it is inferred from Figure 3.1, it plays a central role in the system. It does not only interact with the remaining components, but also drives their behaviour by setting their functional parameters. In the REMPARK system the mobile gateway is implemented using a commercial smartphone and specifically developed software.
- **Drug pump**: It is intended to administer an appropriate drug dose depending on the patient's movement parameters (in REMPARK project, this part was only considered, for a feasibility demonstration and was treated from a theoretical point of view with a laboratory proof of concept).
- **Auditory cueing system (ACS)**: This component provides an auditory pattern intended to assist the patient in the case of abnormal movement patterns (such as FOG or shortened stride length).

3.2.2 The Mid-Term Level

The Mid-term Level (also called the Level 2), refers to the actuation in the medium-long term and constitutes a closed semi-automatic loop, as it will allow the intervention of medical professionals. The system is able to send data to the server, allowing the patient's neurologist to regularly follow the evolution of the patient's disease in a more effective manner, as well as being able to take better informed decisions about the adjustment of the pharmacological treatment of the patient using the Disease Management System (DMS). The system is also able to generate alerts, according personalized thresholds, related to the status and evolution of the disease.

Data on the server can be automatically included in an Electronic Health Record (EHR). This way, the correct intelligent data treatment will help to evaluate and predict the evolution of the disease of a particular patient.

Below, the sub-systems responsible for the storage of the data generated at Immediate Level as well as for its high-level analysis are listed. The results of the high-level analysis are available to both the patient and the neurologist or clinician. The clinician can also interact with the system by changing/updating the medication profile of the patient.

- **Mobile gateway**: At this level, it is responsible for sending the data generated at the first level to the REMPARK server and to get the feedback test to be presented to the patient.
- **REMPARK server**: It is in charge of storing the data provided by the mobile gateway and of making this data available to the Rule Engine, described below.
- **Rule Engine**: This component filters the raw data stored in the server and makes it available to the patient and the caregiver using an appropriate interface. Through this interface, both the PD patient and the clinician/neurologist can be interconnected. The patient is able to access the personal record stored in the server. The neurologist can access the personal records of the supervised patients and eventually update their treatment plan based on this information.

3.3 Definition of the REMPARK System Main Characteristics

This section includes a description of the main characteristics of the system. Some characteristics were included according considerations already contained in the proposal of the project and after the analysis of the results obtained from the answers provided by potential users of the REMPARK system (patients, neurologists and carers) to specific questionnaires elaborated by the consortium [1, 2]. These will imply concrete requirements and constraints for the system that were considered and included in the final version of the tested system.

The indicated references [1, 2] are publicly available reports on these activities which contain details of the administered questionnaires, the constraints of this administration, obtained answers and their analysis. As it is indicated in [1], there are three main conclusions derived from the answers provided by patients to the REMPARK questionnaire:

- The most prevalent motor disorders among patients are: ON/OFF phenomena, dyskinesia and FOG.
- The most prevalent gait problems among patients are: "reduced walking speed", "small steps" and "shuffling". These problems appear both at home and outdoors, and nearly half of the patients use a strategy to improve them.
- Over 80% of patients are familiar with the use of a mobile phone for making calls and sending SMS. However, they are not so familiar with smartphones based on a touch interface.

Regarding the answers provided by caregivers, as stated in [2], four main conclusions can be derived:

- "Reduced walking speed" and "small steps" are the most clinically relevant symptoms in the mild stages of PD. "Freezing of gait" and "difficulty in turning" are the most clinically relevant symptoms in the moderate stages of PD. In the advanced stages of PD, "falls" arises as a new main clinical occurrence.
- Patients use strategies to improve symptoms mainly in the moderate-advanced stages of PD. However, the strategies that PD patients adopt (mostly the use of a stick) are not considered as useful as verbal cueing (auditory).
- "Reduced walking speed" should be considered a symptom as relevant as "freezing of gait".
- The moderate stages of PD are the best target for the REMPARK system. This means that the system should pay special attention to the multifaceted clinical expression of PD during these stages.

Table 3.1 summarises the requirements for the REMPARK system derived from the feedback got from the final users through the proposed questionnaires.

It is worth noting that first and third requirements in Table 3.1 were already considered in the original goals and specifications of the REMPARK project, but second one was only detected after the evaluation of the questionnaires.

Table 3.1   Technical specifications derived from the user feedback

Requirement Heading	Requirement Description
Symptom detection	The system must be able to detect at least the following symptoms: "reduced walking speed", "small steps", "freezing of gait", "dyskinesia", "bradykinesia" and "falls".
Patient interface	The user interface in the mobile gateway must be operated by a PD patient in any stage of PD.
Symptom mitigation	The system must provide auditory cueing upon detection of "reduced walking speed", "small steps" and "freezing of gait" symptoms.

Concerning the symptom detection requirements of the system, the motor status of the patients will be evaluated based on the following parameters:

- Tremor
- Bradykinesia
- Freezing of gait (FOG)
- Stride length
- Gait speed
- Fall indicator
- Dyskinesia

The final conclusion on ON/OFF state is achieved by a real time execution of the included algorithms and an evaluation of their combination or presence during a given time period.

3.4 Subsystems Specification

In previous sections, main components of the REMPARK system and the processes that enable their interaction have been identified. Based on this, the present section elaborates the main technical specification for each individual component: the sensor module, the ACS Auditory cueing sub-system, the smartphone and the REMPARK platform.

As it will be discussed in Chapter 4, a very important part of REMPARK project was devoted to the gathering and construction of a Database, with the cooperation of a number of patients in order to extract the necessary knowledge for the development and implementation of the corresponding algorithms to be embedded in the processing part of the final sensor module. These embedded algorithms in the sensor module will be able to process and to determine in real time the presence of specific symptoms in a patient. It must be clarified here that the presented and mentioned sensors hereafter (the wrist and the waist ones) have been used for data gathering and construction of the Database, but after discussion and final conclusions of REMPARK, the wrist sensor was finally not considered in the pilots, where only the sensor module located in the waist of the patient was finally used.

3.4.1 Sensor Module

The sensor module is the subsystem in charge of determining in real time if a specific indicator (of a concrete symptom) appears in the PD patient movement pattern. The detection of the indicators is obtained after a local process of the data acquired by the sensors included in the module. These indicators are sent at regular time intervals to the mobile gateway subsystem, that will thereafter decide the action to be performed based on their values.

The type and number of sensors to be used depend on the specific movement disorders to be detected. As it has been stated in the previous section, the movement disorders to be addressed by the REMPARK system are the following: Tremor, Stride length, Gait speed, Bradykinesia, Falls, Dyskinesia and FOG.

- Tremor consists in an involuntary, rhythmical, forward and backwards movements of a body part, which are Caused by the rapid alternating contraction and relaxation of muscles. For three out of every four people who develop Parkinson's Disease (PD), the disease begins with a trembling or shaking in one of the hands. It can also appear in the feet, face or jaw. This rhythmic movement of the extremities in PD patients has a frequency typically between 4–7 Hz. Therefore, in order to detect this movement disorder, the REMPARK system should include a sensor unit to be placed in the wrist. This unit should contain a triaxial accelerometer and a data processing unit. In order to provide an indicator whose value is related to the presence of tremor in the hand the data processing unit should be able to carry out sensor readings with a frequency of at least 20 Hz.
- The determination of indicators related to stride length, gait speed and bradykinesia are interrelated, since the gait speed is given by the number of strides in a specified time window, and once stride length and gait speed are determined it is possible to derive an indicator for bradykinesia. Even if the stride length has been determined in the past using gyroscopes placed in the leg [3], it has been demonstrated [4] that a single triaxial

accelerometer placed on the patient's waist can be used to accurately measure stride length, gait speed, bradykinesia and falls.

- Dyskinesia is a medication side effect experienced by patients with Parkinson's Disease and shown in the form of involuntary movements. There are many papers in the literature that manage to successfully detect dyskinesia using inertial sensors. The most prominent research work in this area is performed by Manson et al. [5] which uses a triaxial accelerometer placed on the shoulder. The analysis is based on the comparison of a signal's characteristic with the severity of dyskinesia assessed by Abnormal Involuntary Movement Scale (AIMS) while the patient performs the following activities: sitting, talking, writing, drinking, preparing food, eating and walking. Characterizing the signal consists of obtaining the average value of the accelerations in the range of 1 to 3 Hz, obtained through a band-pass filter. This characteristic of the signal is demonstrated to be correlated with the value of the AIMS score. The study included 26 patients with Parkinson's Disease who were monitored for 20 minutes. Since a triaxial accelerometer placed on the waist will be used to measure stride length, gait speed and falls, the data provided by this sensor unit will also be used for detecting dyskinesia.

- Regarding the detection of FOG episodes, the first work that analysed the relationship between the frequency content of the signals generated by triaxial accelerometers and the presence of a FOG episode was presented in [6]. Two accelerometers were placed in the ankles of both healthy people and PD patients. It was observed that normal walking has a principal frequency around 2 Hz, while FOG episodes are characterized by a principal frequency in the range of 6–8 Hz. A second study was carried out in [7] with PD patients. In this study, a single triaxial accelerometer was placed on the patient's shin, and the signal was measured on the vertical axis (parallel to the leg). It was found that FOG episodes caused the appearance of frequency components in the range 3–8 Hz. This study was continued in [8, 9], where the analysis presented in [7] was extended to accelerometers placed in the patient's thigh and waist. Regarding the results obtained when the sensor was placed on the waist, it was found that they were similar and even better than those obtained when using other body locations. Therefore, the REMPARK system will use the waist accelerometer to derive an indicator related to the rising of a FOG episode. In order to cope with the different measurements that this unit has to carry out the minimum sampling frequency should be 40 Hz.

From the previous explanation, a decision was initially taken and the sensor module to be used in the REMPARK system was constituted by two sensor units. One of these sensor units must be placed on the patient's wrist and the other on the patient's waist. These sensor units must contain, in principle, a triaxial accelerometer and a data processing unit. However, in order to provide additional contextual information to the acceleration data, it is considered convenient to include sensors too providing information related to the speed (gyroscope) and the orientation (compass). This information may improve the sensitivity and specificity of the measurements related to the movement disorder indicators to be generated by the REMPARK system.

As previously indicated, the sensor module has to communicate using a wireless link, such as Bluetooth, with the mobile gateway of the REMPARK system, in order to make the measurements accessible to both the patient and the caregiver. Therefore, in order to simplify the overall system and to make it more user-friendly it is envisioned that the sensor units placed in the waist and the wrist should establish a communication using the same wireless communication protocol.

It is important to bear in mind that during the construction of the already mentioned REMPARK database the raw data captured by the sensors will not be processed by the sensor units, since the goal of this project phase is to get enough representative data in order to develop efficient processing and detection algorithms. Thus, the sensor module should have some amount of local storage in order to save the sensor recordings. This local memory unit should be easily accessible in order to retrieve the data once the experiments with patients are finished.

Another important aspect to be considered is that the sensor module has to be operated by people with no technical knowledge. Hence, it has to offer a very simple procedure to start/stop its operation and to analyse its status at any time (preferably using visual information). Additionally, it has to be worn during extended periods of time (approximately eight hours during the experiments). This poses a constraint to the physical features (weight and size) of the sensor module, to the materials that have to be in contact with the patient's skin.

According to the above considerations, Table 3.2 summarizes some of the main requirements for the sensor module derived from operational features to be exhibited by this subsystem.

Table 3.2   Some important technical requirements for the sensor module

Requirement Heading	Requirement Description
Structure	The sensor module must contain a sensor unit placed in the patient's waist and a sensor unit placed in the patient's wrist.
Size – waist sensor unit	The dimensions of the waist sensor unit must be smaller than 150×70×30 mm.
Size – wrist sensor unit	The dimensions of the wrist sensor unit must be smaller than 80×70×30 mm.
Weight – waist and wrist sensor units	The weight of the sensor units must be low (around 200 g. for the waist unit and 150 g. for the wrist one)
Battery capacity	The battery on the sensor units must permit a normal continuous operation for at least 8 hours.
Operation – waist and wrist sensor units	The sensor units must be turned on/off using a single button.
User interface	The sensor unit must use a single led to display its state.
Communication	The sensor must be able to establish a wireless link with the mobile gateway. This requirement is only for the operative part and not used during the database construction phase.
Communication – additional specs	The waist and wrist sensors must be able to send to the mobile gateway its battery status. The waist sensor must be able to send to the mobile gateway data containing indicators for the following movement patterns: stride length, gait speed, bradykinesia, falls, dyskinesia and FOG.
Communication – security	The data sent by the waist and wrist sensor units to the mobile gateway must be encrypted.
Sensors – waist sensor unit	The waist sensor unit must contain a triaxial accelerometer, a triaxial gyroscope and a compass.
Sensors – wrist sensor	The wrist sensor unit must contain a triaxial accelerometer.
Data sampling rate	The waist sensor unit must be able to sample data from the sensors with a frequency of at least 40 Hz.
	The wrist sensor unit must be able to sample data from the sensors with a frequency of at least 20 Hz.
	It must be noted that during the Database construction, used frequency was higher (200 Hz for the waist sensor and 80 Hz for the wrist sensor).
Data processing	The waist sensor unit must contain a data processing unit able to calculate indicators for the following movement patterns: stride length, gait speed, bradykinesia, falls, dyskinesia and FOG.
	The wrist sensor unit must contain a data processing unit able to calculate indicators for the following movement patterns: tremor.
Comfort	The parts of the sensor unit in contact with the patient's skin must be constructed with a biocompatible material.

(*Continued*)

Table 3.2 Continued

Requirement Heading	Requirement Description
Battery certification	The Li-ion batteries used in the sensor module must have test certificate according to standard UL 1642.
Battery charger certification	The battery chargers used for the sensor module must have a test certificate demonstrating compliance with IEC 60950.

3.4.2 Auditory Cueing Subsystem

This section describes the REMPARK Auditory Cueing System (ACS) system for gait-cueing and FOG intervention. This system intends to be a self-calibrating/adaptive gait guidance system for helping PD patients in real time, during their daily activities.

PD patients are usually affected by symptoms associated with a reduced motor performance and gait disturbances that affect their ability to walk independently and safely [10]. One of the key problems of PD patient's gait is the particular difficulty with the internal regulation of stride length, reflected as an inability to generate sufficient stride length, even though the control of cadence (or step rate) is intact and can be easily modulated under a variety of conditions [11, 12]. Some patients may increase the stepping frequency to compensate the reduced stride length [12–14]. With the disease's progression, other episodic gait disturbances can appear, e.g. start hesitation and FOG episodes [12]. Festination of gait, characterized by small and rapid steps, is also a common symptom of advanced PD [15]. By applying adequate external regulatory rhythmic stimulations, movements' speed and amplitude can be modified [16], so that gait performance is improved, even under complex environments [9, 12, 17, 18].

The ACS systems would therefore be able to adapt the rhythm of cueing to the specific needs of each patient and each situation, being activated and deactivated at the adequate times, taking into account the real time feedback sent by inertial sensors (i.e. relative to the gait's quality), as well as the feedback provided by the patient (i.e. a subjective feedback related with the quality of cueing that is being applied).

3.4.2.1 Gait guidance ACS functional description

The cueing system is intended to be used every day, during daily activities. The smartphone (seen at this point as a control unit) will be responsible for the control of the auditory cueing actuator, and will be able to program the cue parameters taking into consideration the specific needs of the patient in each situation, in a self-adaptive way (Figure 3.3).

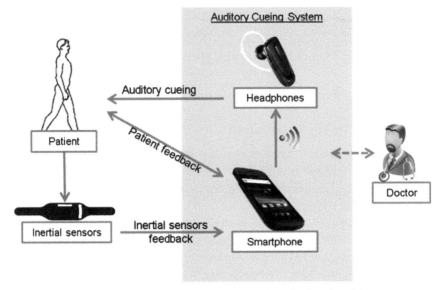

Figure 3.3   Simplified representation of the ACS functionalities.

The system must be mainly controlled in a self-adaptive way and should not require user intervention to trigger or adapt a cueing response. Thus, it is required that the system operates in real time, together with the feedback provided by inertial sensors.

Since continuous auditory cueing might not be acceptable for PD patients [9] the system will provide the cueing automatically only when required, i.e. when the patient is walking and a context of impaired performance or an inefficient walking pattern is perceived (based on the available REMPARK sensor subsystem measures).

In concrete, the final parameters and symptoms considered for triggering of the ACS system are the presence of a bradykinetic walk and FOG episodes, suffered by the patient (see Table 3.3).

The detection of gait impairments is performed through a real-time comparison with the baseline data characterizing the normal range of walking parameters values for a specific individual patient.

It must be clear that, according to the previous explanation, there exists a very close relationship between stride length, speed of walking, cadence and bradykinetic walk. Hence, changes in these parameters, characterizing a bradykinetic walk, must be referred to baseline measurements of a patient [19].

Table 3.3   Parameters/Measures considered by the ACS

Parameter/Measure	Description
Bradykinesia	Slow movements, slow walk
FOG episodes	Transient period in which gait is halted.

This means that even if a low stride length is detected, this parameter can be considered normal if the baseline walking speed is also low.

The determination of "normal values" needs to be done carefully, so that meaningful values are considered for impairments detection. To get this information, some initial tests (called here the calibration sessions) can be performed with each patient before starting to use the REMPARK system. The therapist/doctor must be the responsible for conducting these tests, in order to accurately control the protocols and the results. The REMPARK system will also be able to detect the FOG episodes, which are also considered as an input for the ACS subsystem.

Based on the detection of the triggering symptoms and their comparison with the baseline values, the ACS automatically activates or deactivates the cueing. Also given these measures, the cueing rates must be automatically adjusted to the specific needs of the patient in each situation. This autonomy requires a constant evaluation of the effects of cueing on the patient, and the automatic adjustment of cueing rates according to the patient's response. The requirements related to the specific situations are detailed on Table 3.4.

The ACS will provide the cueing in the form of sounds. Once the patient hears these sounds, he/she will try to synchronize the rhythm of steps with the rate of the sounds provided. Sound beats can be single (i.e. pace one step per stride) or double (i.e. pace both steps per stride). According to [20], gait is more effectively modulated when both footfalls are paced, which means that a double beat must be used. Specific voice recordings of alerts and information must be produced when the patient is not able to synchronize with the rhythm of cues. These alerts must help the patient to synchronize with cueing rate and ask for some patient's feedback.

The ACS implemented and used in REMPARK (see additional information in Chapter 7) is comprised of two physical components – headphones and the smartphone – that will interact (directly or indirectly) with other subsystems. The smartphone will be able to decide when and how to actuate in each specific situation; headphones are placed on the patient's ear and will produce the sounds acting as cueing. The above Figure 3.3 is presenting this architecture.

Table 3.4   General functional requirements of the ACS system

Requirement Heading	Requirement Description
Cueing type	ACS must provide cueing automatically, in a self-adaptive, non-continuous way, taking into account the specific needs of the patient in each situation.
Operation mode	Cueing must operate only when the patient is walking.
System interaction	ACS must be able to work in real time, together with the feedback provided by sensors.
Functional requirements	ACS must be able to detect a bradykinetic gait based on walking speed, stride length and the occurrence of FOG episodes, as measured by movement sensors.
Stimulus type	ACS must provide stimulus in the form of sounds.
Sound rhythm	Sounds must pace both left and right footfalls.
Alerts	ACS must be able to provide voice recordings with alerts and instructions, when required.
Configurability	ACS must be able to program the activation and deactivation of sound stimulus at the adequate times/situations.
Adaptation	ACS must be able to automatically adapt the rhythm of cues (sound stimulus) to the specific needs of each patient in each situation.
Evaluation and Rating	ACS must be able to constantly evaluate the effect of cueing on the patient and to enable the patient to rate the cueing session.

A headset (ear-set) was selected to be included in the auditory cueing system. Ear-sets are usually small and aesthetic and, as they are placed in a single ear, they enable the patient not only to hear the auditory cueing but also to perceive the auditory stimulus from the surrounding environment.

The auditory stimulus will be provided in the form of metronome sounds. Since different rhythms of stimulus are going to be produced, the easiest way to apply cueing is to use metronome sounds, because its rate is easier to be adapted than, e.g., the music.

Since the headset will produce the sounds streamed by the smartphone, it must therefore be compatible with Advanced Audio Distribution Profile (A2DP). This profile defines how audio can be streamed from one device to another over a Bluetooth connection. A2DP is designed to transfer a unidirectional 2-channel stereo audio stream.

Alternatively, audio can be produced by a non-A2DP device, by channelling the audio to the stream that carries phone call audio to the headset. This option was, also, considered as an option.

3.4.2.2 Need of an adaptive ACS

When an external rhythmic auditory pacing is applied, the patient tries to couple his/her footfalls with the beats, which helps to normalize the walking pace (cadence) [20, 21]. As stride length and stride frequency tend to change as a function of walking speed, this adaptation of cadence might offer a strategy to indirectly influence the stride length regulation in these patients, while controlling for walking speed [19].

The effects of auditory cueing may, however, differ for patients with or without FOG, referred as "freezers" and "non-freezers", respectively [18]. It is well known that cueing rates higher than baseline walking rate are contraindicated for freezers. Based on the study by [22], it is recommended the use of a lower rate setting for freezers and an increase of up to 10% for non-freezers. In both cases, the latest goal of training with the ACS is to maximize stride length while walking at as fast a cadence as possible. If these two variables are increased during training, then walking speed is also increased [16].

The cueing rate settings need, then, a careful consideration, i.e. the intended speed of walking and the quality of movements, and the different responses to the cues by different subgroups of patients suffering from PD. Given these conditions, a set of requirements related with the auditory cueing rate must be carefully considered:

- Preference
- Safety
- Walking rhythm
- Different modalities
- Adaptation to the user

As the person can dictate his/her own preferred walking rhythm, the perception of this baseline rhythm is crucial, so that REMPARK ACS can be the most fitted and comfortable as possible to each situation. The comfortable walking rate is used as a basis to establish the auditory cueing tempo being provided if it is necessary.

The baseline rhythm can be perceived when gait is considered normal (without evident motor symptoms). After the person has started to walk a few strides, it is possible to estimate this rhythm (at least 3 strides are required to reach a steady-state walking in healthy young subjects [23]). When this estimation is not possible, the normal cadence at the preferred speed, as calculated during the calibration session, can be used. When setting the cueing rates, it is important to guarantee that the applied rates will never compromise the safety of the person, both during dual task and walk alone.

Therefore, gait cannot be stimulated beyond the person's limits and the stimulus cannot provide a negative effect on the walking pattern of the individual. A too high cueing rate, for example, may lead to small steps (low stride length) at a high cadence, which is an inefficient walking pattern.

As PD patients face different problems and intensities of problems considering the different stages of the disease and medication states (i.e. ON/OFF state and freezer/non-freezer classification) [24, 25], the ACS system must enable the introduction of different settings/configurations for different patients.

The self-adaptive cueing system will have the possibility to be adjusted/ modified in several ways by the patient and by the doctor/therapist. The ACS system must offer the following characteristics regarding user's interaction: configurability to the specific needs of different patients, configuration of different parameters done by caregivers (rate, modality, baseline, rate change …), voluntary deactivation, adjustment of the volume and the tone.

3.4.3 Drug Delivery Pump Considerations

The development of a wireless communication module for subcutaneous infusion pumps of liquid medicines opens an opportunity for real-time control of the applied dose. A system of closed-loop control would be possible for the case of some diseases in which the result of treatment can be measured by a sensor, i.e. the drug infusion could be adjusted to the needs of the patient in real time.

This idea has been explored in the case of diabetes, in which the glucose level in plasma can be measured accurately and thus this information may be used to control the basal dose of insulin to be delivered via a subcutaneous pump infusion. From this hypothesis, various feedback systems that include automatic and semi-automatic (with a doctor in the decision loop) feedback have been successfully developed, in which insulin infusion is controlled according to the level of glucose in the blood. In contrast, drugs for Parkinson's treatment are often given using a pre-established scheduling and dosage. However, regardless of the route of administration, intermittent delivery of the programmed dosage produces fluctuations in the level of the drug in plasma (known as 'peaks and valleys') that produce variations in the effects of the drug in the central nervous system.

The amount of drug in each administration and the time between administrations is calculated so that the peaks of plasma do not exceed the upper therapy limit (which may produce unwanted side effects) and that the valleys do not fall below the lower therapy limit (which may cause the therapeutic effect to disappear). Unfortunately, it is not always possible to maintain drug levels in plasma within the therapeutic range. What is more, as the disease progresses, the goal of staying in the therapeutic range is increasingly difficult to achieve and therefore, the time intervals between administration of drug become shorter and the doses higher. Contradictory to this, such doses and time intervals seldom completely control symptoms and therefore, the OFF periods and dyskinesias occur eventually.

Recently, it has been hypothesized that if the drug levels are maintained within the therapeutic range, control of symptoms associated with Parkinson's Disease will be improved substantially [26].

To this end, therapeutic solutions are being developed, aimed at eliminating the programming/intermittent administration of medication and replacing it by a continuous, controlled administration. It is precisely for this reason that subcutaneous administration of dopaminergic drugs was made possible by the development of infusion pumps that maintain drug levels in plasma in the desired range for as long as possible.

The purpose of using these pumps is to maintain the level of drug in plasma within the desired range, and, at least, change the dose on a schedule based on time (different doses depending on the time of day, assuming the needs of the patient). The development of "automatic" infusion pumps, able to adjust the drug dose based on the patient's symptoms will be welcomed both by patients and by doctors as a breakthrough in controlling the symptoms of Parkinson's Disease.

Apomorphine is an agonist dopaminergic with a very short average life, which is administered subcutaneously. There is extensive experience in the use of apomorphine for the treatment of Parkinson's Disease, including its administration in the form of bolus or flow for the release of "lock states" and the corresponding basal treatment.

In order to address the symptoms in real time with apomorphine, a subcutaneous infusion pump that can adapt to them is required, that is, it should receive orders from a device capable of monitoring symptoms.

Even if drug infusion pumps were not tested with patients within the framework of the REMPARK project, one of the project goals was to test if the REMPARK system may be useful to create an automatic or semi-automatic feedback loop that will improve the treatment of PD patients. For this purpose, a commercial apomorphine infusion pump was used in the REMPARK system. The pump was modified with the necessary electronic components so that it was able to communicate with the mobile gateway so as to receive commands related to the drug dose to be administered. The pump should be able to operate both in bolus and in infusion mode. Additionally, the electronic system added to the pump should permit a manual user interaction.

In REMPARK project, the organization of a specific pilot action, with real patients, using a drug delivery system control was discarded because it was out of the real possibilities of the project activity. Alternatively, it was decided to consider this effective possibility, preparing all the requirements and conditions from a technical and functional point of view. This activity was based on an already developed experience by some of the participants [27].

3.4.4 The smartphone Technical Requirements

The smartphone is at the core of the REMPARK BAN, receiving data from sensors and sending instructions to actuators. Furthermore, it serves as a gateway between the BAN and the server, by receiving settings and preferences from the server and sending the data gathered in the BAN. In consequence, the smartphone is a critical device in the BAN, as without it the system would be unusable. Figure 3.4 shows the connections of the smartphone on the REMPARK system, as well as the directions of the information exchanged.

We split the requirements of the smartphone into hardware and software. The hardware requirements are mainly related to the characteristics of the smartphone including connectivity. It must provide battery life, as well as other usability requirements.

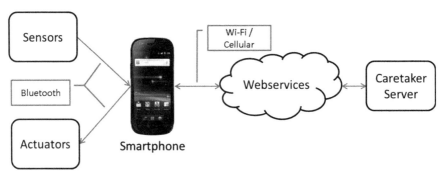

Figure 3.4   Smartphone connectivity and information flows.

Connectivity-wise, the smartphone should be able to form the BAN itself, connecting to all sensors and actuators using Bluetooth protocol. To connect to the caretaker server, it will need an Internet access enabled. Here, cellular connectivity is mandatory (GMS or UMTS) and Wi-Fi will be used whenever possible. Although most smartphones in the market today already come with all these radio interfaces, we need to ensure that they are present and available; for instance, we need to ensure that the smartphone has a SIM card with a data plan enabled, so that cellular data is available. As this is a critical component of the system, we need to ensure that it is always on, or at least minimize its downtime. To this end, it is critical to have a suitable battery that will last for the whole day while the patient is away from a power source. We expect the system to last at least 8 hours working on battery power alone, as this is the maximum expected time a person will be away from a power source.

With regards to the usability of the smartphone, we expect it to be big enough and have enough resolution so that older adults with some visual difficulties can use it. Furthermore, the device needs a touch screen interface to provide direct interaction and a more reduced learning curve.

As a result of a review of relevant literature, and interviews with doctors and relatives, we identified major changes on PD affected people's fine motor skills, changes that makes it very hard for them to do precise movements both on ON and OFF stages. Hence, we need a screen large enough to accommodate buttons as big as necessary in order to offer them good interaction experience.

- On/OFF state, they have their fine motor movements mostly conditioned by Bradykinesia and rigidity. Causing slower movements and loss of agility thus precise movements become very difficult. Possibly they can do some precise selections but with great effort. Whereas tremor although being generally a rest tremor, sometimes can still be active while on movement, and thus makes a precise movement even harder.
- While on ON state even when most PD symptoms are gone, on later stages of the disease, PD affected people have dyskinesia which can be impairing as the other symptoms, since it make them do involuntary movements and thus making it very difficult to do precise selections as well.

Moreover, with the aim of interacting with a touch screen, a user does not need to apply much pressure to activate it, which is also beneficial to PD affected people since they also have lack of strength and are much more comfortable with smooth gestures.

In order to ensure these preconditions and to guarantee that we can extend the smartphone functionality in any way we require it to have an open-source operating system and main drivers, so that we can access, change and compile them to any software version we need.

Concerning the smartphone related software aspects, they will consist of distinct components:

- **Server Communication:** The smartphone will need to exchange information with the REMPARK Server.
- **End-user Applications:** The smartphone will provide a number of applications related to the management of the disease or the REMPARK system.
- **Data Processing and Event Detector:** The smartphone will analyse the data received from the sensors and will raise specific events.
- **Sensor Communication:** The smartphone will gather information from waist sensor.
- **Actuator Communication:** The smartphone will control the actuators upon receiving orders from the Server, an input from the user or after processing the data gathered by the sensors.

In regards of the Server Communication, the smartphone must guarantee the communication with the REMPARK server, establishing a communication channel with the caretaker server whenever needed, and use encryption methods that must ensure:

- **Data integrity:** both the smartphone and the server must guarantee that the information they receive was correctly transmitted and unaltered.
- **Data security:** the mechanism must ensure that no one else can have access to the information being transmitted, as well as ensure that no one can alter the information while it is being transmitted.
- **Authentication:** both the smartphone and the server will be able to authenticate themselves to each other, thus guaranteeing that information is being transmitted to the right party.
- **Non-repudiation:** tightly coupled to the authentication mechanism, non-repudiation aims at guaranteeing that: (1) every message is signed, so that the sender cannot deny that it had sent the message; (2) no one else can impersonate the sender. This will guarantee that every message is accounted for.

The final set of the smartphone requirements is related to provide a set of applications, both giving users and medical caretaker's tools to better manage this disease. To the users, it provides assistive applications to help them with their daily lives and to the medical caretakers it provides valuable medical information of the state of the disease. These applications range from actuators controllers (e.g. for auditory cueing), assistive applications to help the users with their daily living (e.g. medication reminders) questionnaires to input medical or routine information and prompts to validate specific situations detected when acute events are detected.

The smartphone is also the interface to the caretaker server, so it must provide a way for patients to use the services provided by the server, e.g., check health status, contact the doctor, see appointments. Table 3.5 provides an overview of these requirements.

The REMPARK smartphone must receive and process the data from the sensors. The processing will be achieved through a set of algorithms to assess the status of the patient and control the BAN actuators. These algorithms will be adapted to the particular needs of each user. As such, the smartphone will fine-tune the algorithms by updating its information with the caretaker server. Furthermore, after analysing the data, the smartphone should decide on the action, or event, to trigger. These actions can be:

- to enable an actuator for the interaction with the patient
- to raise an alarm so that medical staff can respond
- to ask some input, or confirmation, to the patient.

The smartphone will also provide an Actuator Communication component to interact with the actuators. The actuator actions must be triggered either by the smartphone in the cases described above, or directly by the server if, for instance the medical staff decides that the patient should receive some input in real time (e.g., a medical questionnaire). Furthermore, when communicating with the actuators, the smartphone must ensure that: (1) messages sent to the actuator are not replicated, so that the actuator does not trigger the same action twice; (2) no messages for the actuator are lost, to guarantee that the actuator really acts; (3) actions of the actuator are personalized for each patient, for instance, the auditory cueing system uses the frequencies that better suit the patient.

Table 3.5 Requirements of the smartphone end-user applications

Requirement Heading	Requirement Description
Questionnaires	The smartphone must enable the user to answer medical questionnaires sent by the doctor.
User answers	The smartphone must enable the user to answer specific prompts to validate alerting detected situations.
User input	The smartphone should enable the user to input routine information such as the time of intake of the medications, quantity and quality of the sleep or other information.
Actuators	The smartphone should enable the user to adjust the behaviour of the auditory cueing.

3.4.5 The REMPARK Platform Architecture and Functionality

The REMPARK platform is composed of all the services that will store and process information, namely, the server and the Rule Engine blocks. While the so-called server will have the role of keeping the database and hosting all the services, the Rule Engine will use these data to make further analysis and store new processed-data. All this information will be used by the medical application to show relevant information to clinical professionals that is why it is got an interaction with the server. Likewise, the Mobile Gateway will also interface the server to store measures and alerts. Figure 3.5 shows an overall architecture view of the main blocks that make up the REMPARK's platform and its main interactions.

An important element of the REMPARK platform is the server, enabling the potential service. There are two main functionalities. The first one deals with measures and the second one with alerts. Both functions are deployed as REMPARK's services in the server and they will provide also methods to insert and extract both measures and alerts of any type. These services will be exposed so that the blocks that interface the Server (i.e. Rule Engine, Medical Application and Mobile Gateway) can use them.

Other identified service is the Patient Service. The Server must provide a way to register new patients and associate them with their information (i.e. measures, alerts) as well as providing some basic information report about them and their associated hospital, if necessary. Finally, it might be useful to have questionnaires' information available and that is why a specific service dealing with this functionality should be implemented. Figure 3.6 shows the services that the server will expose to their interfaces.

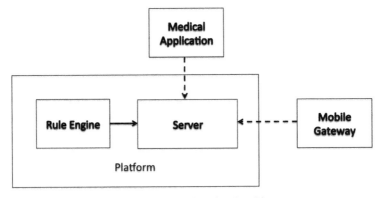

Figure 3.5   Platform functional architecture.

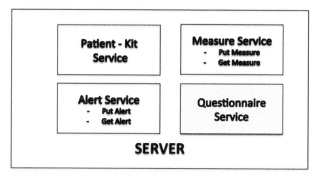

Figure 3.6 Server services.

Table 3.6 summarizes the main technical requirements of the REMPARK server

Table 3.6 Technical requirements for the REMPARK server

Requirement Heading	Requirement Description
Service	Server must expose some public services so that the Mobile Gateway, the Rule Engine and the Professional Application can access the generated data.
Service	Server must provide services that are able to store and extract for each specific patient.
Measures	Server must provide a service to store measures which will be used by the Mobile Gateway, Rule Engine and the Professional's application.
Measures	New measures must be able to be added in a simple fashion.
Getting Measures	Server must provide a service that allows receiving measures and is accessible by the Mobile Gateway, Rule Engine and Professional Application.
Alerts	Server must provide a service to notify about alerts which will be used by the Mobile Gateway, Rule Engine and the Professional's application.
Alerts	New alerts must be able to be added in a simple fashion
Storing Alerts – Mobile Gateway	• Patient has fallen. • No connection with motion sensor. • Emergency button. • Sensor is running out of battery. • Mobile phone is running out of battery.
Storing Alerts – Rule Engine	• Mobile Gateway has lost connection with the platform. • Mobile Gateway has resumed connection with the platform.

3.4.5.1 Important functional parts

As it was indicated in Figure 3.2, there is a set of important functional pieces embedded in the REMPARK server that are crucial for the implementation of its final functionality. They are the Rule Engine sub-system (RE), the Disease Management System (DMS) and the user-Web interface.

3.4.5.1.1 *Rule Engine*

The Rule Engine (RE) is a subsystem which is intended to analyze and process data from sensors at regular time intervals. The main idea is to perform tasks, such as post processing (e.g. filtering and transformation) or data analysis (e.g. verify that the most recent pulse and/or blood pressure measurement was within an acceptable range), at regular time intervals. The tasks and their timing were defined by medical and technical partners along the project.

Four components make up the RE: (1) a timer, (2) a processing service, (3) a task manager and (4) a data manager. The data manager encapsulates the access to the main REMPARK database. It is the main interface to REMPARK system and provides the means to read and write data (i.e. measures and alerts). The task manager will decide when a task needs to be executed at a certain point in time. The processing service is fed with a set of tasks which it executes. A timer is used to (re-)initialize the processing of tasks at regular time intervals.

In order to implement the tasks, it must be specified: (1) how often a particular task needs to be performed (e.g. every minute, every five minutes), (2) the data that the performance of this particular task requires (e.g. pulse data or data from the inertial sensor), (3) how recent the data needs to be (e.g. last fifteen minutes, at least two hours old or just the most recent value) and (4) the task itself (e.g. check whether pulse is within acceptable range).

3.4.5.1.2 *Disease Management System*

The Disease Management System (DMS) is an application for managing the patient's health by the medical staff. Data and medical information are integrated into the DMS to support medical decisions. These decisions have influence on changes in the treatment plan and raising alerts to the medical team who has to take care of it. The DMS for REMPARK was developed as a web application using the .NET platform with C# as the programming language.

The DMS is based on the REMPARK database as the resource for every data coming from the patient. The information is sent through the communication module and written to the database. The DMS will have access to the data as a raw data and to the processed data coming from the RE.

Every user of the DMS will be linked to a specific user profile. Every profile will be exposed to the relevant data according to pre-defined specifications. Four profiles have been defined:

- Supervisor – The main authority in the site (Doctor, nurse or other) that will have access to the management parts of the system. The supervisor can register new users and manage the profiles.
- Doctor – The Doctor is the highest medical authority in the call center/ point of care. Therefore, the doctor will have full access to the treatment plans of the patients related to the specific patients' doctor.
- Nurse – Every nurse will be assigned to several patients who will be monitored by another specific nurse. The supervisor will have the ability to allow one nurse to access another nurse's patient in case of absence of the nurses. This exceptional access will be time limited. A nurse can be assigned also to all patients at the call center.
- Patient – The patient will be related as a DMS user in order to allow the patient access a web interface/web site for watching the information related to him/her. A patient cannot be deleted from the database. Instead, the patient's status can be changed to "Not Active" and it will not be shown in the DMS lists (or report) but will remain in the database.

Important operative tools, included in the DMS are:

- Patient's Record. The patient's record must include the patient details, patient relative details (a contact person in case of emergency), a technical part with documentation of the equipment implemented in the patient's home (ID, IP, etc.) and a treatment plan.
- Treatment Plan. For each patient registered in the REMPARK program a treatment plan is created by the neurologist according to the patient's evaluation and clinical history. The treatment plan will include the necessary exercises, medications, normal ranges for every measurement the patient will have the sensors connected to the Mobile Gateway at the patient's home.
- All of the above will be documented in the DMS. The treatment plan can be viewed by any of the medical team assigned to the patient but only the doctor can change an existing medication treatment plan. In some cases, a nurse could also be able to change the treatment plan. These special cases, if any, will be defined in the Clinical Protocol. A change in the treatment plan can be a result of changes in the patient's health, patient's environment, etc.
- Alerts. The alerts will be raised in two ways.

1) Alerts that were sent by the Rule Engine after processing raw data.
2) Alerts that came directly from the Mobile Gateway without any pre-processing.

- Clinical protocols. The clinical protocols can be described as set of flow charts allowing the neurologists to make decisions according to the collected information. It is actually a set of medical instructions for every situation, change (or lack of change) in the patient status. These protocols can be updated or changed from time to time according to the medical considerations.

Table 3.7 compiles the requirements of the DMS according to the description.

3.4.5.2 Platform technical constraints

Before analysing the different services that must be supported by the server, it is important to estimate important parameters that can compromise the performance of the whole system, such as the maximum storage, the maximum number of simultaneous transactions and the bandwidth.

Just as an example, calculations were made considering the constraints imposed by the pilots organized in REMPARK project. Final technical decisions were:

- Maximum storage requirement of 100 Gbytes, permitting to operate with 60 patients during 1,5 years, considering that a probable transmission rate of 1 Kbyte per minute should be necessary, with the sensor connected during 12 hours a day.

Table 3.7   Compiled requirements for the DMS

Requirement Heading	Requirement Description
BAN data	DMS will store all data from the BAN.
Medical monitoring	DMS will monitor medical aspects of the patient.
Technical monitoring	DMS will monitor technical aspects of the patient regarding the REMPARK project.
Alerts	DMS will raise alerts when information about irregular behaviour or measurement will come from the gateway.
Questionnaires	DMS will allow managing questionnaires.
Treatment plan	The treatment plan will be created in the DMS.
Data integrity	No data could be deleted from the database.
Monitoring interface	The DMS will have a monitoring interface.
Data interface	The DMS will have an interface for showing patient's data.
Website	The DMS will have a patient personal website.
Reports	The DMS will be able to publish reports.

- Concerning the bandwidth, it was estimated that 1 Mbps channel is more than enough for the requirements of REMPARK.
- The maximum number of parallel transactions, considering the previous figures, overcome the necessary characteristics.

3.5 Conclusion

The present chapter has presented and described the architecture, functionality and technical requirements of the REMPARK system, organized along some different sub-systems specifically designed for fulfilling the functional objectives of the proposed project, finally implemented and tested in real piloting experiences.

References

[1] REMPARK Deliverable D1.1 – Questionnaire addressed to patients. Answers and statistical results. Publicly available.

[2] REMPARK Deliverable D1.2 – Questionnaire addressed to doctors/ physiotherapists. Answers and statistical results. Publicly available.

[3] A. Salarian et al., "Gait Assessment in Parkinson's Disease: Toward an Ambulatory System for Long-Term Monitoring," *IEEE Transactions on Biomedical Engineering*, vol. 51, no. 8, pp. 1434–1443, August 2004.

[4] A. Rodríguez-Molinero et al., "Detection of Gait Parameters, Bradykinesia, and Falls in Patients with Parkinson's Disease by Using a Unique Triaxial Accelerometer," *Movement Disorders*, vol. 22, no. S3, p. S646, September 2010.

[5] A. J. Manson et al., "An Ambulatory Dyskinesia Monitor," *J. Neurol Neurosurg Psychiatry*, vol. 68, no. 2, pp. 196–201, February 2000.

[6] J. H. Han, W. J. Lee, T. B. Ahn, B. S. Jeon, and K. S. Park, "Gait Analysis for Freezing Detection in Patients with Movement Disorder Using Three Dimensional Acceleration System," in *Proceedings of 25th IEEE EMBS*, Cancun, 2003, pp. 1863–1865.

[7] S. T. Moore, H. G. MacDougall, and W. G. Ondo, "Ambulatory Monitoring of Freezing of Gait in Parkinson's," vol. 167, no. 2, pp. 340–348, January 2008.

[8] M. Bächlin et al., "Online Detection of Freezing of Gait in Parkinson's Disease Patients: A Performance Characterization," in *Proceedings of the 4th International Conference on Body Area Networks*, Los Angeles, 2009.

[9] M. Bächlin et al., "A wearable system to assist walking of Parkinson's disease patients," *Methods of Information in Medicine*, vol. 49, no. 1, pp. 88–95, January 2010.

[10] M. E. Morris, F. Huxham, J. McGinley, K. Dodd, and R. Iansek, "The biomechanics and motor control of gait in Parkinson disease," *Clinical Biomechanics*, vol. 16, no. 6, pp. 459–470, July 2001.

[11] J. M. Hausdorff, "Gait dynamics in Parkinson's disease: common and distinct behavior among stride length, gait variability, and fractal-like scaling," *Chaos (Woodbury, N.Y.)*, vol. 19, no. 2, pp. 1–14, June 2009.

[12] M. F. del Olmo and J. Cudeiro, "Temporal variability of gait in Parkinson disease: effects of a rehabilitation programme based on rhythmic sound cues," *Parkinsonism & Related Disorders*, vol. 11, no. 1, pp. 25–33, January 2005.

[13] D. Schaafsma et al., "Gait dynamics in Parkinson's disease: relationship to Parkinsonian features, falls and response to levodopa," *Journal of the Neurological Sciences*, vol. 212, pp. 47–53, 2003.

[14] E. van Wegen et al., "The effect of rhythmic somatosensory cueing on gait in patients with Parkinson's disease," *Journal of the Neurological Sciences*, vol. 248, no. 1–2, pp. 210–214, October 2006.

[15] N. Giladi, H. Shabtai, E. Rozenberg, and E. Shabtai, "Gait festination in Parkinson's disease," *Parkinsonism & Related Disorders*, vol. 7, pp. 135–138, 2001.

[16] M. P. Ford, L. A. Malone, I. Nyikos, R. Yelisetty, and C. S. Bickel, "Gait Training With Progressive External Auditory Cueing in Persons With Parkinson's Disease," *Archieves of Physical Medicine and Rehabilitation*, vol. 91, no. 8, pp. 1254–1261, August 2010.

[17] Y. Baram and A. Miller, "Auditory feedback control for improvement of gait in patients with Multiple Sclerosis," *Journal of the Neurological Sciences*, vol. 254, no. 1–2, pp. 90–94, March 2007.

[18] W. Nanhoe-Mahabier et al., "The possible price of auditory cueing: Influence on obstacle avoidance in Parkinson's disease," *Movement Disorders*, vol. 0, no. 0, pp. 1–5, February 2012.

[19] E. Wegen et al., "The effects of visual rhythms and optic flow on stride patterns of patients with Parkinson's disease," *Parkinsonism and Related Disorders*, vol. 12, no. 1, pp. 21–27, June 2006.

[20] M. Roerdink, P. J. Bank, C. Peper, and P. J. Beek, "Walking to the beat of different drums: Practical implications for the use of acoustic rhythms in gait rehabilitation," *Gait & Posture*, vol. 33, no. 4, pp. 690–694, March 2011.

[21] E. Cubo, S. Leurgans, and C. G. Goetz, "Short-term and practice effects of metronome pacing in Parkinson's disease patients with gait freezing while in the 'on' state: randomized single blind evaluation," *Parkinsonism and Related Disorders*, vol. 10, no. 8, pp. 507–510, May 2004.

[22] A. M. Willems et al., "The use of rhythmic auditory cues to influence gait in patients with Parkinson's disease, the differential effect for freezers and non-freezers, an explorative study," *Disability and Rehabilitation*, vol. 28, no. 11, pp. 721–728, June 2006.

[23] B. Najafi, D. Miller, B. Jarrett, and J. Wrobel, "Does footwear type impact the number of steps required to reach gait steady-state?: An innovative look at the impact of foot orthoses on gait initiation," *Gait and Posture*, vol. 32, no. 1, pp. 29–33, May 2010.

[24] G. McIntosh, S. Brown, and R. Rice, "Rhythmic auditory-motor facilitation of gait patterns in patients with Parkinson's disease," *Journal of Neurology, Neurosurgery, and Psychiatry*, vol. 62, pp. 22–26, 1997.

[25] A. Delval, V. Weerdesteyn, J. Duysens, and S. Overeem, "Walking patterns in Parkinson's disease with and without freezing of gait," *Neuroscience*, vol. 182, pp. 217–224, 2011.

[26] D. Deleu, Y. Hanssens, and M. G. Northway, "Subcutaneous apomorphine: an evidence-based review of its use in Parkinson's disease," *Drugs Aging*, vol. 21, no. 11, pp. 687–709, 2004.

[27] A. Rodríguez-Molinero, D. A. Pérez-Martínez, et al. "Remote control of apomorphine infusion rate in Parkinson's disease: Real-time dose variations according to the patients' motor state. A proof of concept". *Parkinsonism Related Disorders* (8), pp. 996–998, 2015.

4

Assessment of Motor Symptoms

**Albert Sama[1], Carlos Pérez[1], Daniel Rodriguez-Martin[1],
Alejandro Rodriguez-Molinero[2], Sheila Alcaine[3], Berta Mestre[3],
Anna Prats[3] and M. Cruz Crespo[3]**

[1]Universitat Politècnica de Catalunya – UPC, CETpD – Technical Research Centre for Dependency Care and Autonomous Living,
Vilanova i la Geltrú (Barcelona), Spain
[2]National University of Galway – NUIG, Galway, Ireland
[3]Centro Médico Teknon – Grupo Hospitalario Quirón, Parkinson Unit, Barcelona, Spain

4.1 Introduction

This chapter describes the methodology developed and used to obtain the elements of the REMPARK system devoted to monitor the related PD motor symptoms, according to the specifications discussed in the previous chapter. The monitoring part basically consists in the waist-worn device and its embedded algorithms for the analysis of the PD patients' movement and gait.

The development of such a device to assess motor symptoms has been divided into several steps that are summarised below:

- First, different questionnaires were administered to professionals and cargiveers to identify some system requirements and the most important symptoms to be monitored. This step has been already partially presented and used for the purpose of Chapter 3. The present chapter describes the details of these questionnaires and the obtained results.
- A methodology was developed to detect these symptoms, which is based on inertial sensors and machine learning techniques. This methodology is presented in Section 4.3.
- A database of signals was collected in order to develop automatic detection methods based on the mentioned methodology. The design of the experiment to collect such data and a summary of the data obtained are also detailed in Section 4.3.

- The algorithmic approach to exploit the database of signals based on machine learning techniques is presented in Section 4.4. Each symptom and parameter detected by an algorithm is described in a different subsection.
- Finally, some conclusions are drawn.

4.2 Decision on the Most Relevant Symptoms to Be Detected and Assessed

During REMPARK project, specific tasks were devoted both to the collection of clinical information and to understand aspects that may be relevant for the REMPARK system. These tasks were carried out during the first months of the project through the administration of two questionnaires addressed to patients and related professionals [1]. Their results were included in the system design to set the technical specification of the system (partially discussed in Chapter 3).

The most relevant issues addressed in the questionnaires administered to professionals are summarised below:

- First, on the salient clinical features of PD in the different stages of the disease. Here, for instance, professionals were requested to indicate the symptoms with higher priority for treatment, to report which monitored symptoms may have a greater impact on daily living, and which are the most frequent symptoms according to a mild, moderate or severe PD.
- Second, information was collected on how the professional expects to be helped by REMPARK system in the therapeutic management of the individuals with PD. "*At which PD stages do you think REMPARK may be useful?*", "*For the improvement of which symptoms do you think REMPARK may be useful?*", are examples of the related questions made to the professionals to acquire valuable information at this level of analysis.
- The third issue addressed consisted on how data had to be organised for an optimal use and updating of the PD treatment in the disease course. At this level professionals were required to express their opinion on the usability of REMPARK in PD patients as a function of symptoms severity.

Furthermore, the interviews helped the consortium to understand the perspective, the expectations and the general attitude of the care service (i.e., clinical professional) against the REMPARK approach. At this regard, emphasis was paid in capturing different perspectives. For this reason, the questionnaires were administered to different kind of professionals belonging to the medical (i.e., neurologists, geriatricians) and technical rehabilitative (i.e., physiotherapist, occupational therapist) areas.

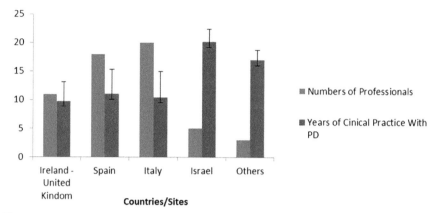

Figure 4.1 Illustration of the number of professional who filled the questionnaire divided according to their country of origin. Years of clinical practice in PD area are also reported in the average, also in this case individually for each country. Vertical bars represent the standard deviation.

4.2.1 Subjects

Some characteristics of the 57 professionals recruited for the administration of the questionnaires are reported in Figure 4.1. The average number of years of clinical expertise with PD patients was 13.7 (SD=4.6). Most of the clinicians were employed in Public Health Service (n=36) while 18 of them were employed in private or "intermediate" health care system.

Across the different countries/sites, it must be noted that the three kinds of professional who took part in the study were neurologists, geriatricians and physiotherapists. The majority of participating professionals were neurologists (n=23), followed by physiotherapists (n=22) and, then, by geriatricians (n=10).

4.2.2 Questionnaire

The characteristics of the questionnaire including the formulation of the items and methodology for answering questions were developed through a continuous consultation between the four medical partners participating in REMPARK: Centro Medico Teknon (Spain), Fondazione Santa Lucia (Italy), Maccabi Healthcare Services (Israel) and the National University of Ireland at Galway (Ireland).

The questionnaire is composed of three main sections.

- An initial social-demographic section in which the participant is required to provide personal information regarding, for instance, own specialty, the country of origin and the years of clinical experience with PD patients.

- A second central section that addresses clinical issues related to PD. This is the section in which participants indicate the clinical relevance of PD symptoms according to the three disease stages (i.e., mild, moderate and advanced). Questions such as the following ones were posed:

  - *What do you consider are the three most characteristic motor symptoms of this phase?*
  - *What do you consider are the 3 motor symptoms that interfere the most, with the quality of life of people with Parkinson's at this stage of the disease?*
  - *What do you consider are the three priorities to treat symptoms at this stage of the disease?*

The professional was asked to answer by ticking a square box in a mixed multiple alternative forced choice paradigm. In fact, for most questions, if the professional feels that the right answer does not fit with the proposed alternatives he can tick the square box corresponding to "other" and, then, is allowed to better specify his response.

- The third and final section of the questionnaire aimed at investigating the potential utility of REMPARK system for the clinical management of PD, as it is perceived by professionals. Questions like the following ones were proposed here:

  - *Do you consider that a system such as REMPARK would be useful to improve motor problems of your people with Parkinson's?*
  - *In your clinical practice, do you consider that a system such as REMPARK would be a useful system for monitoring motor problems of your people with Parkinson's?*

In the case of the professional expressing a positive judgment about REMPARK utility by ticking the "yes" box, he is required to indicate both the PD stage for which REMPARK could be better applicable (i.e., mild, moderate or advanced stages) and which PD symptoms would benefit from REMPARK utilisation. Also for these questions the professional has to respond by ticking a square box in a mixed multiple alternative forced choice paradigm.

4.2.3 Results

4.2.3.1 Analysis of the correlation between responses on clinical questions

A first item analysis was performed to investigate the coherence of the professionals' response relating the clinical answers. More specifically, the participant is required to evaluate the clinical relevance of a PD symptom by indicating:

- The three most characteristic motor symptoms of PD.
- The three motor symptoms that interfere more with the quality of life of people with PD.
- The three symptoms that have priority for treatment.

All three questions were individually addressed for mild, moderate and advanced PD stage. The three questions are apparently related since it can be reasonably posited that the most characteristic symptoms of PD have a great probability to be those symptoms that interfere with quality of life and, furthermore, those for which a treatment is imperative. Therefore, from the statistical point of view, the existence of a significant correlation between the responses on these items could be a parameter to verify the reliability of responses themselves.

In order to examine the correlation between the professionals' responses on the above three items/questions, Pearson' r statistic was performed. For the purpose of these analyses, in order to quantify the relative weight of each symptom, the responses were classified according to a Likert-type scale ranging from 0 to 3, resulting in most of the correlations analysed being significant. In these cases, the r value ranged from 0.27 to 0.72 being $>$ 0.40 in about 67% of all cases. The significance of the correlation was only approached in one case relating to the analysis that involved the "Difficulty in Turning" symptom (i.e., the correlation between the score attributed to "most characteristic symptoms" and to the "priority for treatment items").

Therefore, the correlation analyses, by confirming the existence of a significant relationship between the professionals' response, indicate a global coherence of the responses themselves.

4.2.3.2 Investigation of the clinical relevance of the motor symptoms in the three PD phases (i.e., mild, moderate and advanced)

The clinical relevance of the PD symptoms as reported by professionals was investigated by means of descriptive analyses firstly without taking into consideration the particular country/site where the data were collected and, then, in a second step, individually for each country/site. This was made in order to have both a general view on data and to evidence possible differences as a function of the country/site the professional belongs to.

For the purpose of these analyses, a unique score was computed by collapsing the score attributed to each individual item in the three clinical questions mentioned in the previous sub-section:

An index of clinical relevance for each symptom was, thus, computed by averaging the score attributed by professionals to that symptom in the three questions. For instance, the index of clinical relevance for "small steps" was represented by averaging the scores attributed to it in questions *1*), *2*) and *3*). Also in this case, for the purpose of these analyses, in order to quantify the relative weight of each symptom a score of 3 was attributed to the symptoms the professional indicated as first, a score of 2 was attributed to a symptom indicated as second, a score of 1 was given to the symptom indicated as third and, finally, a score of 0 was attributed to the symptoms not included in the first three symptoms list.

As showed in Figure 4.2, independently from the countries/sites where data were collected, the analysis of the index of clinical relevance computed for PD motor symptoms evidence some differences according to the PD stage considered.

In fact, in the mild stage (represented by blue columns in the figure), "reduced walking speed", "small steps", "difficulty in turning" and "shuffle" were the four symptoms with the higher index of clinical relevance with and index value of 1.74, 1.24, 0.68 and 0.66, respectively.

In the moderate stage (represented by red columns in the figure), the most clinically relevant symptoms were "freezing of gait", "difficulty in turning",

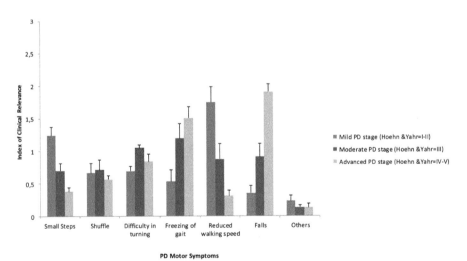

Figure 4.2   Indices of clinical relevance for each PD motor symptoms examined referred to the mild (blue columns), moderate (red columns) and advanced (green columns) PD stages. Vertical bars represent standard errors.

"falls" and "reduced walking speed" with an index value of 1.19, 1.05, 0.91 and 0.87, respectively.

Finally, as for the advanced stage (represented by green columns in the figure), the symptoms with a higher index of clinical relevance were "falls", "freezing of gait", "difficulty in turning" and "shuffle" with an index value of 1.91, 1.50, 0.84 and 0.56, respectively.

According to the majority of professionals, patients with PD adopt some specific strategies to improve gait difficulties particularly in the moderate-advanced stages of the disease. More specifically, more than 80% (n=48) and about 91% (n=52) of professionals indicates that PD patients use strategies to improve gait in the moderate and advanced phases, respectively, compared to the 33% who report this behaviour in the mild stages of the disease. *The strategies more frequently adopted by patients in the moderate-advance disease stages would be stick use, verbal cueing, attention focus on walking and steps counting.* These strategies have been indicated by about 79% (n=38) of professionals for the moderate stage and by the 80% (n=42) of them for the advanced stage.

However, according to professionals, the most useful strategies to be adopted in the moderate-advanced PD would be the use of verbal cueing for about 32% (n=18), steps counting for about 16% (n=9) and stick use for about 14% (n=8) of them. As for the advanced stage, also in this case the majority of professionals indicated verbal cueing as the most useful strategy to be adopted (about 26% of responses; n=15) followed by stick (about 23% of responses; n=13) and metronome (about 12% of responses; n=7) use.

4.2.3.3 Analysis of REMPARK utility for PD patients

A large majority of professionals considered REMPARK potentially useful for PD management. More specifically, about 96% of professionals (n=55) judged REMPARK a useful system for symptoms improvement and 93% (n=53) considered REMPARK potentially useful for symptoms monitoring. The REMPARK utility for both symptoms improvement and monitoring, was perceived by professionals for the intervention in the moderate stages (49% and 40% of responses, respectively) and at a lesser extent in the mild (about 21% and 23% of responses, respectively) and advanced (about 9% and 19% of responses) stages. However, the presence of missing data (n=13) related to the lack of responses on the specification of the PD phase for which the application of REMPARK system would be more useful, reduces the reliability of these findings.

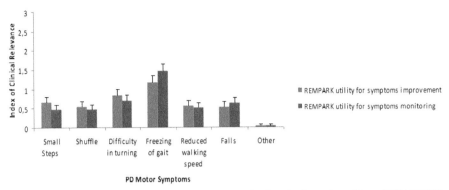

Figure 4.3   Subjective judgment expressed by professionals about the utility of REMPARK system for both improvement and monitoring of motor symptoms. Vertical bars represent standard errors.

In order to quantify the professionals' judgements about which symptoms REMPARK system would be more useful, in terms of both improvement and monitoring, the professionals' responses were classified on a Likert-type scale ranging from 0 to 3, where 0 represents the minimum value assigned to the effect of REMPARK system on a specified symptom and, conversely, 3 indicates the highest value.

As Figure 4.3 illustrates, a substantial coherence is noted about the symptoms that would better benefit from the application of the REMPARK system in terms of symptoms improvement and monitoring. Indeed, according to professionals' opinions, the symptoms on which the REMPARK system would have a greater positive impact would be "freezing of gait" and "reduced walking speed".

4.2.4 Discussion and Conclusive Remarks

A first critical issue the questionnaires are focussed on is the understanding of the salient clinical features of PD in the different stages of the disease. The answers should also give valuable information about clinicians' expectations on the REMPARK utility and usability in the clinical management of PD.

A first preliminary comment has to be devoted to the reliability of the professional's answers. Regarding this aspect, some indicators such as the absence of missing data on content questions as well as a substantial coherence of responses on clinical questions suggests that the questionnaires have been filled correctly and congruently.

The descriptive and inferential statistics applied to data allowed us to evidence some main points of interest.

- First, as expected, different symptoms achieve a clinical relevance and require a therapeutic intervention as a function of different PD phase considered and, thus, in particular in the mild stage of PD "**reduced walking speed**" and "**small steps**", were the most clinically relevant symptoms whereas in the moderate stages "**freezing of gait**" and "**difficulty in turning**" appear to be more important clinical signs. Moreover, in the advanced stages, "**falls**" arises as a new main clinical occurrence. As mentioned above, this finding is expected on the basis of the neurological characteristics of PD [2]. Indeed, PD is a neurodegenerative disease that progressively affects different motor and non-motor brain circuitries with a related modification of both the qualitative and quantitative (i.e., severity) clinical features of the disease [3–5]. An interesting aspect to be remarked is that according to professionals, PD patients use strategies to improve symptoms mainly in the moderate-advanced stages, rather than in the mild ones. This finding is obviously expected on the basis of the greater impact that symptoms severity progressively exerts on daily living. However, according to professionals, *the strategies that PD patients seem to adopt for improving their gait difficulties do not appear to be the most useful*. In particular, professionals judge to be useful strategies to be adopted in both the moderate and advanced stages the verbal cueing whereas it seems that PD patients tend to use stick more frequently.

Furthermore, the analysis of questionnaires outlines a substantial convergence of the professionals' clinical judgements between the different countries/sites for mild and advanced stages of the disease. However, it should be noted that the same judgements appear to be more heterogeneous when applied to moderate PD stages. A possible interpretation of this heterogeneity is related to the objective difficulty to clinically define the moderate stages in respect to mild and advanced ones.

- A second main point evidenced by the analysis is that *REMPARK system is perceived by professionals as a potentially useful instrument for the management and treatment of PD*. This is particularly observed in the moderate stages of the disease. As a matter of fact, the majority of professionals indicated the moderate phases of the disease as the best target phase for REMPARK. In this regard, it should be noted here that, as previously discussed, the moderate stage of the disease is the stage for which the judgment on the clinical relevance of symptoms is more heterogeneous. This provides a clear indication for REMPARK. Indeed, the multifaceted clinical expression of PD during this phase should be taken into account carefully to develop a functional system.

- Finally, as for the advanced stage, the symptoms with a higher index of clinical relevance were "falls", "freezing of gait", "difficulty in turning" and "shuffle" with an index value of 1.91, 1.50, 0.84 and 0.56, respectively.

In conclusion, from the analysis and considerations done, the ***REMPARK system appears to be perceived particularly useful to be applied for both monitoring and improving PD symptoms in the moderate-advanced stages of the disease***.

REMPARK system might be a useful and well accepted instrument for the therapeutic management of PD. Additionally, there exists evidence that PD patients spontaneously adopt strategies to improve gait disorders by using external aids.

4.3 Methodology and Database to Monitor Motor Symptoms

This section presents the implemented methodology in REMPARK system for the detection of the main motor symptoms discussed above. As this methodology will be based on an artificial intelligence approach, it is necessary the construction of a specific database for the required knowledge extraction.

4.3.1 An Artificial Intelligence Approach and the Need of Relevant Data

The main objective of the REMPARK project is to obtain a system capable of assessing PD motor and non-motor symptoms. This is a clinical goal that is intended to be solved through technological solutions.

Firstly, REMPARK system involves inertial sensors to monitor PD motor symptoms given their nature. Since motor symptoms affect movement, inertial sensors capable of measuring such movement are used to automatically detect these symptoms. Secondly, the techniques used to determine their presence come from the Artificial Intelligence (AI) field; more concretely, machine learning techniques are well-known to provide high accuracies in these tasks. In this case, REMPARK work was focused on supervised learning methods.

Supervised learning techniques for classification tasks are mathematical and statistical methods that are capable of recognising patterns to associate them with specific classes. These methods require sets of labelled data in which patterns and their corresponding class labels are given. In the case of REMPARK, inertial signals labelled with the presence of symptoms are needed. In consequence, a specific data capture is required to gather such **labelled datasets**.

Machine learning techniques require the maximum amount of data and the most variability in them in order to properly generalise an automatic detection from them. In addition, labels must be as accurate as possible. Through these data, highly accurate models capable of automatically classifying the patterns can be obtained. In consequence, REMPARK envisaged the construction of a database of labelled inertial signals from 90 PD patients from 4 different countries.

It must be taken into account that the usage of supervised learning techniques creates some restrictions into the algorithmic development, which will be carried out after the database collection. Data collection must follow a strict protocol designed according to clinical restrictions in order to capture the required variety of PD symptoms in different severities. The statistical representability of the data will enable supervised learning techniques to extract the embedded knowledge and, thus, precisely detect the presence of symptoms into the signals provided by inertial sensors.

4.3.2 Protocol for the Database Construction

The data for the database were collected in the most homogeneous possible way, and under the best conditions to ensure good enough generalization capabilities. It is a very relevant task since the validity of the REMPARK system for assessing a patient's motor status relies on the quality of the data in the database.

A specific clinical study was designed and carried out in order to collect the database. It was a multicentre international study that was conducted in four European settings: Centro Médico Teknon (Spain), National University of Ireland, Galway (Ireland), Fondazione Santa Lucia (Italy) and Maccabi (Israel).

The primary objectives of the study that collected the data were:

- To obtain a database of properly identified inertial signals, which will allow the training of processing **algorithms for motor phase** detection (ON/OFF) in PD patients.

- To obtain a database of properly identified inertial signals, which will allow to train processing **algorithms for motor symptoms** detection in PD patients.
  - To obtain identified inertial signals of hand tremor.
  - To obtain identified inertial signals of freezing of gait.
  - To obtain identified inertial signals of bradykinesia of the lower and upper limbs.
  - To obtain identified inertial signals of dyskinesia of the trunk and limbs.
- To obtain a database of properly identified inertial signals corresponding to movements and activities that can be mistaken for PD **motor symptoms** (potential **false positives**).
- To obtain a database of properly identified inertial signals corresponding to **gait parameters**.
  - To obtain identified inertial signals of gait speed.
  - To obtain identified inertial signals of step/stride length.
- To obtain a database of properly identified inertial signals corresponding to movements and activities that can be mistaken for **falls** (potential **false positives**).

The reference population was that formed by Parkinson's patients with moderate to severe disease and motor symptoms (Hoehn and Yahr greater or equal to 2.5 including ON/OFF phases, FOG or dyskinesia). The total number of recruited patients was 92, distributed among the clinical centres (26 in Spain, 16 in Ireland, 24 in Italy and 26 in Israel). A convenience sampling stratified by symptoms was conducted, keeping desired minimum proportions of patients with different motor symptoms. At least 50% of the sample were set to have ON/OFF motor fluctuations, with the OFF state characterized by bradykinesia. Furthermore, at least 25% of the sample had to present FOG episodes and, finally, at least 25% of the sample was set to present dyskinesia (at least 15% will present trunk dyskinesia).

The inclusion criteria for these patients were:

- to have a clinical diagnosis of Idiopathic Parkinson's Disease according to the UK Parkinson's Disease Society Brain Bank [6]
- disease in moderate-severe phase (Hoehn and Yahr greater or equal to 2.5) with motor fluctuations with bradykinesia, FOG and/or dyskinesia
- aged between 50 and 75 years and willing to participate in the study and wanting to co-operate in all its parts

- accepting the performance regulations and procedures provided by the researchers.

Patients fulfilling the following criteria were excluded from the study:

- other health problems that hamper physical activity
- rheumatologic, neuromuscular, respiratory, cardiologic problems or significant pain
- carriers of implanted electronic devices: cardiac pacemaker, implantable automatic defibrillator . . .
- patients receiving continuous therapy using intestinal duodopa or apomorphine
- patients who have received deep cerebral stimulation therapy (neurosurgical procedure)
- chronic consumption of psychotropic drugs and/or alcohol
- known mental disease, such as dementia, according to clinical criteria -DSM-IV-TR and MMSE score $\leq$ 24 or neuropsychiatric disorders
- patients who are participating in another clinical trial
- patients unable to fully understand the potential risks and benefits of the study and give informed consent
- subjects who are unable or unwilling to cooperate with study procedures.

The data capture was conducted in two visits. The first visit comprised both the inclusion and basal visit, where the inclusion criteria were confirmed, and initial clinical and socio-demographic data of the patient were gathered. The second visit was devoted to the experimental procedures, where the maximum number of physical signs related to the disease were recorded using the inertial sensors and standard methods. This visit had two types of experiments that will happen in interleaved manner (according to the symptoms that the patient may present in each moment).

- The first type of experiments consists in short controlled tests, where the patient was asked to perform certain activities, with the aim of capturing specific motor symptoms (bradykinesia, dyskinesia, freezing of gait, etc.). These tests were closely controlled, using video recording as a gold standard.
- The second kind of experiments that took place in this visit involved monitoring of the free activity of the patient, and recording the natural symptoms that he/she spontaneously may present. This monitoring lasted hours, and the activity and symptoms were electronically recorded by trained observers (using a tabled and specific software).

OFF		ON	
Controlled experiments	Free monitoring	Controlled experiments	Free monitoring

Figure 4.4   Design of the experimental visit.

The two types of experiments took place alternatively, according to the motor state and the symptoms that the patient presented. That is to say that when the patient was in an OFF phase, the specific controlled tests for the OFF symptoms were conducted (e.g., FOG) and the remaining OFF state time was used for monitoring their free-natural activity. Similarly, when the patient entered the ON phase, some specific short tests for capturing ON symptoms (e.g., dyskinesia) were performed, with the rest of the time devoted to monitoring the free natural activity of the patient in this state. Figure 4.4 summarizes the experiment done.

All participants were trained to follow the specific study procedures, according to a common protocol that was the same for all the study sites. Patients also received specific training for recognizing their own OFF state. For this purpose, specific videos showing other patients in ON and OFF states were displayed, and detailed explanations on symptoms defining the OFF state were provided.

The investigators received a 3-day training session, comprising theoretical sessions including guidelines and instructions of all the instruments and questions of the Case Report Form (CRF), and practical sessions with pretended patients who behaved according a number of pre-established situations which served an example of the most relevant cases. The entire experimental test was performed at least twice by all the researchers, and every researcher conducted an example free monitoring session of at least 60 minutes.

During this training session, investigators were also trained into the usage of the designed labelling tools:

- Labelling for the controlled experiments was done once the data capturing with the patients had finished. Researchers were trained into the usage of the tool, that allowed, first, the synchronisation of the inertial signals with the videos, and, second, the labelling of the different symptoms that were listed in the objective of the study.
- Labelling of the free-monitoring experiment was done in-situ by the investigators. A tablet with a specific application were used by them. This application enabled the annotation of the different symptoms at the same time that the inertial signals were captured.

Finally, the Principal Investigator, or his designee, in accordance with institutional policy, obtained an Informed Consent that was reviewed and accepted by the Ethics Committee. A written consent form bearing the full name, date and signature of the patient and the local investigator were obtained from each patient. The signed Informed Consent constitutes a confidential document and therefore was archived in the study binder. A copy of the consent was also given to the patient.

The inertial signals captured during this data collection phase were obtained through two sensors: a waist sensor and a wrist sensor. The waist sensor was worn inside a pocket within a neoprene belt. The wrist sensor was worn through a strap. Figures 4.5 and 4.6 present both devices.

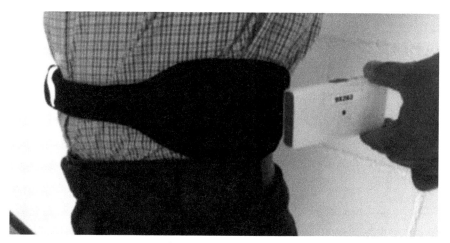

Figure 4.5   Waist sensor.

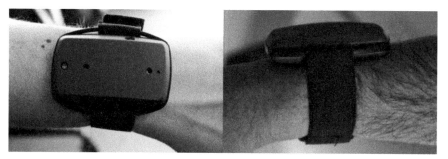

Figure 4.6   Wrist sensor.

4.3.3 Gathered Database Description

The following data were included in the database for each patient:

1. Socio-demographic data: Age, Sex, Educational level and Marital status
2. Parkinson's Disease related information:

   - Parkinson's severity, as measured by Hoehn and Yahr scale
   - Date of symptoms' onset
   - Date of PD diagnosis
   - Motor section of UPDRS in OFF phase
   - Motor section of UPDRS in ON phase
   - Information on OFF periods characteristics and duration (UPDRS motor complications section – motor fluctuations)
   - Information on FOG presence, characteristics and duration (FOG questionnaire)
   - Information on dyskinesia presence, characteristics and duration (UPDRS motor complications section – dyskinesia)
   - List of treatments

3. Co-morbidity related information. Cognitive status: Mini-Mental State Examination, Test of Attentional Performance and List of conditions.
4. Inertial signals labelled according to the following motor symptoms:

   - Motor phases. Signals were labelled among the three following options: ON, OFF and Intermediate state.
   - Dyskinesia severity and location. Dyskinesia is a side effect of medication, not a PD symptom, and signals were labelled according to the following modalities: Weak Trunk dyskinesia, Weak Foot/Leg dyskinesia, Weak Hand/Arm dyskinesia, Weak Head dyskinesia, Strong Trunk dyskinesia, Strong Foot/Leg dyskinesia, Strong Hand/Arm dyskinesia, Strong Head dyskinesia.
   - Bradykinetic gait (presence/absence). This symptom describes a difficulty to walk and slow gait, including small steps, shuffling and difficulty to turn.
   - FOG type. Episodes were labelled according to their type: Start Hesitation FOG, Straight Line FOG, Turning FOG, Tight FOG, Destination FOG.
   - Tremor location and severity: Modalities labelled were: Right Hand/Arm tremor, Right Foot/Leg tremor, Trunk tremor, Left Hand/Arm tremor, Left Foot/Leg tremor.

5. Inertial signals labelled according to body postures and activities: sitting, standing, walking, going upstairs, going downstairs, elevator (down),

elevator (up), walk with FOG, carrying delicate object, carrying heavy object, lying, jumping, running

6. Other information from inertial signals:

- Falls
- Walking aids: scooter, walking stick, walker, crutch, crutches, lean on furniture, tripod walking stick

Organization of the complete set of data from each patient is summarised in Figure 4.7.

Database contains clinical data from 92 participants with idiopathic Parkinson's Disease. Regarding the sociodemographic data, as Table 4.1 shows, fifty-six of them are male (60.9%) and 36 (39.1%) are female. The average age of the participants is 68 (SD 7.9). Seventy-four patients are married or live with a couple (80.5%), 10 (10.8%) single or divorced, and 8 (8.7%) widow.

All the participants in the database construction are patients with moderate disease, having a Hoehn and Yahr scale of 2 or more. Average Hoehn and Yahr score is 3 (IQR 0.5). The average time from diagnosis of the disease was 10.5 years (SD 12.2). Eighty-nine patients (96.7%) of the database have OFF periods, according to the "motor complications" section of the UPDRS. Sixty-eight (73.9%) had predictable off periods, 54 (58.7%) had unpredictable

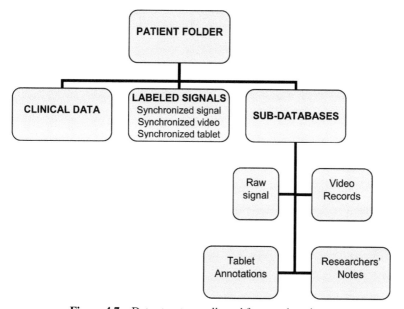

Figure 4.7 Data structure collected from each patient.

Table 4.1 Sociodemographic data

Age (Mean ± SD)	68 (7.9)
Gender	
Female	36 (39.1%)
Male	56 (60.9%)
Marital Status	
Single	5 (5.4%)
Married/partner	74 (80.5%)
Widowed	8 (8.7%)
Separated/divorced	5 (5.4%)

off periods and 33 (35.9%) had sudden off periods. Most of them spend less than a quarter of the day in OFF. 34.7% of the total declared to spend more than 50% of the daytime in off.

Sixty-four patients (69.6%) present some degree of dyskinesia, being non-disabling dyskinesia in 54.3% and non-painful in 78.3%. Twenty-five patients have dystonia (27.2%). Only 5 patients have a 0 score in the FOG-Q, meaning that the rest of them present some gait problems.

The database contains inertial signals properly identified and labelled according to Parkinson's motor symptoms and body postures and activities. In total, the database contains 406 hours of inertial signals. Information on the motor status of the participant is available for 346 hours of inertial signals of the database (see Table 4.2).

A total of 175 hours of motor symptoms are recorded and identified in the database (including bradykinesia, dyskinesia, FOG and tremor). Thirty-two of them correspond to inertial signals labelled against a video record gold standard, and the rest correspond to inertial signals which have been labelled using the real-time notations of an observer (tablet-PC annotations). Table 4.3 summarizes the time (hours) of symptoms recorded and labelled in the database.

Table 4.4 shows the amount of motor symptoms (bradykinesia, dyskinesia, FOG and tremor) recorded in each motor phase (ON, OFF or "Intermediate"), according to the gold standard used (video records vs tablet-PC annotations).

Table 4.2 Recorded time of the different motor periods

Motor Phase	Time Recorded (hours)
ON	163
OFF	111
Intermediate	72

Table 4.3 Video recording duration per symptom in the database

	Dyskinesia	Bradykinesia	FOG	Tremor	TOTAL
Video	8,10 h	15,78 h	2,45 h	5,60 h	31,92 h
Tablet-PC	62,82 h	31,82 h	2,96 h	45,82 h	143,43 h
Total	70,93 h	47,60 h	5,41 h	51,42 h	175,36 h

Table 4.4 Summary of motor symptoms per motor phase

	On (minutes)		Off (minutes)		Intermediate (minutes)		Motor Phase Not Available		TOTAL (minutes)	
	Video	Tablet	Video	Tablet	Video	Tablet	Video	Tablet	Video	Tablet
Dyskinesia	355	2500	28	431	18	712	85	126	486	3769
Bradykinesia	50	122	790	1394	25	308	81	85	947	1909
FOG	21	33	113	76	7	36	6	34	147	178
Tremor	94	789	224	1200	15	680	2	80	336	2749
Total	520	3444	1155	3100	65	1737	174	325		

4.4 Algorithmic Approach and Results

Once presented the complete scenario of the objective motor symptoms to be studied from a clinical perspective, it is necessary to propose an algorithmic methodological approach to bring all this knowledge closer to the achievement of machine-learning classifiers (algorithms) for the motor states monitoring tasks.

The methodological proposal to estimate the motor state (ON or OFF periods) of a PD patient wearing an inertial device is based on the use of a hierarchical system. In a first level, the system permits to put in context the patient's activity and a second level is in charge of the detection of the symptomatology of interest. The hierarchical system uses the output of the detection algorithms in this second level for the assessment of the patient's motor status. Some details are presented in Figure 4.8.

The contextualization of the patient's activity and posture is very important because the evaluation of the different PD symptoms is related to the activity developed by the patient. Thus, evaluation of bradykinesia will only be performed when the patient is walking, since it is during self-executing activities when this symptom is clearly manifested. In this way, the inertial signals from the primary accelerometer sensor are analysed using temporary windowing with a set of algorithms that determines if the patient is walking and, if so, signals are analysed to determine the presence of Bradykinesia. This strategy is applied in a similar way for Dyskinesia, where the detection is performed only in the case that the contextualization algorithms determine that

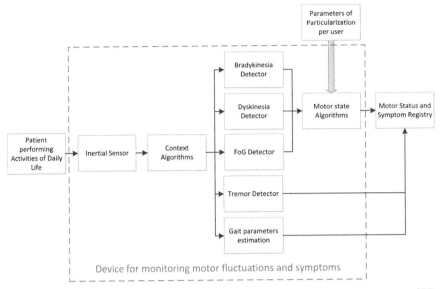

Figure 4.8   Outline of the structure defined for the algorithms for detecting symptoms of PD.

the patient is not walking during the windowed analysis, since it is considered that the gait hides dyskinetic movements.

The information obtained from the detectors of bradykinesia and dyskinesia, together with the FOG detector results will be analysed through a set of additional algorithms, which will determine the final motor state of the PD patient (ON/OFF states). At this stage, as it is indicated in Figure 4.8, it is necessary to use some personal parameters of the patient (basically obtained from previous medical history information).

The development of the different detection algorithms, corresponding to each considered symptom was done independently, taking as a starting point some relevant papers published so far, analysing them, exploiting the acquired signals contained in the REMPARK database and trying to improve, when possible, the previous published results. As it has been already mentioned, the methodology used is based on a machine-learning approach, mainly using supervised learning techniques. The available database described in the above Section 4.3 was used for this purpose.

In the machine-learning area, it is a common practice to divide the Database into different sub-sets. One sub-set is strictly used for algorithmic training purposes and other sub-set is only used for testing. Related works performed in REMPARK project used this approach and the results of the different techniques were evaluated with the patients' data not used in the training process of the supervised learning algorithms.

The signals labelled according to the symptoms listed in Table 4.4 were used to train the different supervised learning models. It must be considered that all the included algorithms process accelerometer measurements sampled at 40 Hz. In addition to the symptoms listed in Table 4.4 (dyskinesia, bradykinesia, FOG, and tremor), there is another subsection devoted to describe the estimation of gait parameters. A description of the algorithmic work done is presented in the following subsections.

4.4.1 Dyskinesia Detection Algorithm

A processing method based on a frequency analysis of the signal was implemented and used for Dyskinesia detection. The method considers the power spectrum of the concrete band between 1 and 4 Hz for the detection of Dyskinesia, provided that information corresponding to higher frequency band (from 8 to 20 Hz) could correspond to false positives such as walking or climbing stairs. Additionally, a number of conditions are added in order to allow a better contextualization of the patient's movement and consequently, to improve the specificity of the algorithm. The algorithm has been subdivided into two steps (detailed below), one at window level and one at the minute level.

- In the first step (at window level), the evaluation of the presence of Dyskinesia is done through the analysis of three separate frequency bands:
    - **Dyskinesia Band**: A high spectral power density in this band is a clear indication that the patient is suffering Dyskinesia, although it may also mean that the patient is walking or climbing stairs. This band is covering from 0.68 Hz to 4 Hz.
    - **Non-dyskinetic band**: It is considered that this band covers from 8 to 20 Hz. This frequency band allows to discriminate if an increase of spectral power in the band of Dyskinesia is due to the appearance of a Dyskinesia or because the patient is walking (or doing similar activities).
    - **Postural transition band**: This is the band from 0 to 0.68 Hz. The posture transition is a very common action and involves very low frequencies that can generate harmonics in the Dyskinesia band, which may provoke false positives.
- The detection of Dyskinesia, based on frequency band analysis allows us to know, in a given time window, whether or not the patient has

Table 4.5   Dyskinesia algorithm results

Type of Choreic Dyskinesia		Num. of Patients with This Type of Choreic Dyskinesia	Equal Weight per Minute		
Severity	Body Part		Specificity (%)	Sensitivity (%)	Total Minutes
Weak	Trunk	16	95	78	953
Strong	Trunk	4	95	100	895
Weak	No-trunk	32	95	39	1110
Strong	No-trunk	7	95	90	917

Dyskinesia. However, Dyskinesia is a symptom that is repeated over time for many minutes, this fact can be used to minimize, using an aggregation process, the presence of false positives. This method of aggregation allows us to examine the appearance of Dyskinesia in several consecutive windows over time by performing an aggregation of the output of each window providing a unique output in a given time slot (it has been considered a period of 1 minute in the implemented algorithms).

This signal processing method was applied to the database of signals by using a leave-one-patient-out scheme on the minute-basis output of the algorithm.

The main reference for this work is [7] and a summary of the results obtained is provided in Table 4.5. The results were obtained with REMPARK's database of video-labelled signals from 92 PD patients.

Specificities and sensitivities are provided for each different type of dyskinesia. With regards to the most important in the clinical sense, i.e. the strong trunk dyskinesia, the algorithm achieves a specificity of 95% and a sensitivity of 100%.

4.4.2 Bradykinesia Detection Algorithm

Bradykinesia appears when plasma's dopamine level is low, seriously complicating the general mobility of the patient and, in particular, causing changes and compromising the way of walking.

The analysis done has been based on the gait cycles characterization. The detector algorithm identifies, on the one hand, the strides that the patient is currently carrying out and, on the other hand, characterizes these gait cycles, allowing the analysis of Bradykinesia through specific characteristics that correlate with the presence of a pathological alteration.

In the analysis of the database, it was concluded that the most important feature that helps to diagnose the occurrence of Bradykinesia is the *fluidity of movement* when walking. Fluidity is a **highly subjective variable** that is not currently objectively measurable. When a patient presents a very low movement fluidity, the probability of manifesting Bradykinesia is very high. This principle allows to generate an algorithmic approach that objectively measures characteristics associated with the fluidity of the movement and, furthermore, enables the detection of the symptom based on comparative thresholds, which determines when the symptom is present.

The bradykinesia detection algorithm is structured using a three-stage scheme (see Figure 4.9). In the first one, the contextualization of the movement is realized, detecting if the patient is walking or not. In the second block, a process of recognition and identification of strides/steps is performed, and the last block performs the analysis of the characteristics of strides/steps that may be representative of the occurrence of Bradykinesia.

The walking detection done at the first stage is performed through a process of pattern recognition from the obtained accelerometer signals. A binary-classification procedure has been approached to detect gait through the use of Support Vector Machines (SVM). The input of the SVM consists of a group of features, which are extracted from a temporary window of signal obtained from the accelerometer. The training set for the SVM was generated through the windows obtained from the signals corresponding to a group of 10 patients, which were acquired from a previously obtained database and its associated gold standard. It should be noted that these patients were not used, later, in the validation group of the final implemented algorithm. The most relevant considered characteristics for the detection of the gait are the power spectra in the three spatial axes, for the bands from 0.1 to 3 Hz and from 0.1 to 10 Hz.

The step detection process is launched when the SVM walking detection has been positive on a given window. This step detection is carried out by recognizing the biomechanical characteristics of walking in the acceleration measurements taken from the sensor (located in the waist).

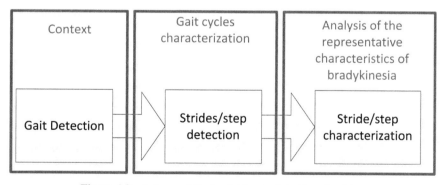

Figure 4.9   Scheme of the bradykinesia detection algorithm.

Table 4.6   Bradykinesia algorithm results

Specificity	Sensitivity	PPV	NPV
81%	88%	89%	84%

The interest of the analysis focuses on the strides (two consecutive steps of each feet) and their characterization, in order to represent the fluidity of the patient's movement. Several statistical markers have been studied and evaluated for this purpose, bearing in mind that the best marker is the one maximizing the separation between the presence and absence of Bradykinesia.

Additionally, it must be considered that the states (Bradykinesia presence or not) are very dependent on the user and, therefore, the threshold that correctly separates the states of a particular patient may have a value different from the threshold of another patient. The main reference of this work is [8] and a summary of results is provided in Table 4.6. These results were obtained by analysing the data from the 92 PD patients who participated in the database construction.

4.4.3 Tremor Detection Algorithm

Tremor was evaluated by analysing the signals provided by the wrist sensor included in the REMPARK system. A frequency analysis of the signals was performed, permitting the extraction of several characteristic features in order to determine the presence of the symptom. The process is based on a SVM model.

The signal processing approach is divided into two different phases: the window level in which tremor is recognized based on short duration signals and the meta-analysis level that aggregates several window detections.

1. At the window level, frequency related features of the signals are used, because this is one of the most common methods. We observed that frequencies in the band from 4 to 6 Hz appear when Parkinsonian tremor is present, and these frequencies are not observed when this type of tremor is absent (which is in agreement with current literature).

   Given the main frequency behaviour of this sort of tremor, it could be theoretically detected only using frequency characteristics. However, a list of other features has been used in the literature for this purpose (for instance, Fast Fourier Transform (FFT), Peak frequency and its amplitude, Entropy of signal, Sum of first, second and third harmonic . . .). In order to measure the impact of non-frequency features in the accurate detection of tremor, two approaches were defined. On the one hand, a first method only used frequency features while, on the other hand, the second approach also included non-frequency features that were previously used in the literature (see reference [9] for additional details).

Both approaches are composed of two phases in order to determine if tremor is present in a certain time window:

- Feature extraction phase. Features are defined depending on the used approach: frequency features alone or combined with those mentioned above. Frequency features from three axes were obtained, and their amplitudes were summed up without taking into account the amplitude of the zero-frequency harmonic. Thus, dependence on the sensor's orientation is avoided. From this, the previously described features were acquired.
- Learning phase. An SVM classifier is trained to distinguish tremor and non-tremor windows based on the chosen feature set.

2. At the meta-analysis level, since it is very important to minimize the resources needed for tremor detection, time windows must be as short as possible (i.e. about few seconds). However, short windows are likely to produce false positives (e.g. a single segment with tremor surrounded by non-tremor segments) since short movement may be confused with tremor (e.g. teeth brushing). Thus, a meta-analysis is added in order to enhance the reliability of the proposed approaches.

The employed meta-analysis method considers the algorithm's outputs in a set of several consecutive windows covering a period of several seconds. These outputs are aggregated into a value representing the probability of having tremor in the corresponding period. This period is considered as tremor if the probability is greater than a certain threshold.

Following the common procedure in the field, the database was split into three non-overlapping sub-datasets: training, holdout and test. A SVM classifier was trained to distinguish tremor and non-tremor windows, using the training sub-dataset. The final evaluation was done on the test dataset and indicates the performance of the developed algorithms.

In total two feature sets (i.e. only frequency features vs. commonly employed features) and two SVM kernels (i.e. linear vs. Radial Basis Function (RBF)) were evaluated.

The main reference for the tremor algorithmic approach is [9] and a summary of results is provided in Table 4.7, where each column represents a different learning model: "RBF+ Freq." corresponds to a SVM with RBF kernel and frequency features, "Lin+Freq." corresponds to a SVM with linear kernel and frequency features, "RBF+All" corresponds to a SVM with

Table 4.7   Tremor algorithm results as presented in [9]

	RBF+Freq.	Lin.+Freq.	RBF+All	Lin.+All
Sensitivity (holdout)	100,00%	100,00%	100,00%	90,00%
Specificity (holdout)	98,50%	99,50%	99,30%	97,20%
Data Usage (holdout)	57,70%	41,10%	42,00%	82,10%
Sensitivity (test)	97,30%	91,00%	98,10%	92,10%
Specificity (test)	96,90%	99,00%	98,60%	97,50%
Data Usage (test)	55,50%	40,80%	42,00%	79,90%
Geometric Mean (test)	97,10%	94,90%	98,40%	94,80%
Accuracy (test)	96,90%	98,60%	98,60%	97,30%

RBF kernel and both frequency and temporal features, and, finally, "Lin+All" corresponds to a SVM with linear kernel and both temporal and frequency features. These results were obtained by training the method with data from 18 patients and validating it with data from 74 patients.

4.4.4 Freezing of Gait (FOG) Detection Algorithm

Freezing of Gait (FOG) is a widely studied and evaluated symptom from the point of view of automatic detection methodology, since it is one of the most disabling symptoms for the patients and one of the most difficult to be evaluated by clinicians.

As it is clear from the current literature, detection techniques for the laboratory setting are highly developed at the moment, and they have had relatively high success rates. However, many problems arise when we tried to apply these methods to the daily living activities, because many false positive appeared due to the new situations and movements appearing under non-controlled scenarios.

In the literature, it has been identified a frequency band on the acceleration signals from the lower limbs of PD patients associated with FOG episodes and ranged between 3 and 8 Hz. In consulted work, a freezing index is defined based on the ratio of the square of the spectral power of these frequencies associated with the freezing band to the square of the spectral power of the frequency band corresponding to the act of walking, (between 0.5 and 3 Hz).

Since FOG mainly occurs when starting, during or at the end of the gait, it is essential to contextualize the patient's activity through a gait detection algorithm. We can take advantage of the gait detector based on the SVM presented in the Subsection 4.4.2 (for Bradykinesia detection). Some points must be considered:

- Contextualization was implemented in the sense that positive FOG detection is validated when the algorithm detected that the patient is walking or has been walking for the last 5 seconds.
- The onset of gait is a complex detection since, in the case of a *posteriori* detection, the condition that validates the detection may never occur because probably the patient would experiment a fall, or because the patient's FOG lasts longer than the imposed temporary condition.
- In addition to adding the validation condition of 5 seconds walking to the formulation, some significant detectable events were considered to know when a patient is rising from the sit position (transition from sitting to standing). This action is very important for the contextualization of FOG since a large number of episodes occurs some moments after the patient is performing this action and try to walk. With this objective, the postural transition band was used for the detection of these events.

In summary, this symptom is detected based on a set of both temporal and frequency features, similarly to the tremor detection algorithm, although the presented contextualisation is added. The main reference for the algorithmic approach is [10] and a summary of results is provided in Table 4.8, where each column represents a different learning model: "RBF freq." corresponds to a SVM with RBF kernel and frequency features, "Linear Freq." corresponds to a SVM with linear kernel and frequency features, "RBF All" corresponds to a SVM with RBF kernel and both frequency and temporal features, and, finally,

Table 4.8   Freezing of Gait algorithm results

Kernel	RBF	Linear	RBF	Linear
Features	Freq.	Freq.	All	All
Sensitivity (train)	100,00%	92,30%	100,00%	92,30%
Specificity (train)	100,00%	100,00%	100,00%	100,00%
Data Usage (train)	69,60%	89,10%	90,60%	98,60%
Geometric Mean (train)	100,00%	96,10%	100,00%	96,10%
Accuracy (train)	100,00%	98,70%	100,00%	98,50%
True Positives	9	8	9	12
False Positives	0	0	0	0
True Negatives	55	82	65	65
False Negatives	1	1	1	1
Sensitivity (test)	90,00%	88,90%	90,00%	92,30%
Specificity (test)	100,00%	100,00%	100,00%	100,00%
Data Usage (test)	82,30%	91,90%	94,90%	98,70%
Geometric Mean (test)	94,90%	94,30%	94,90%	96,10%
Accuracy (test)	98,50%	98,90%	98,70%	98,70%

"Linear All" corresponds to a SVM with linear kernel and both temporal and frequency features.

In this case, a subset of the whole REMPARK database was used. More specifically, these results were obtained by using signals from 15 patients as a training set and the resulting model being evaluated with signals from other 5 patients. Training was done with patients who had FOG episodes. Validation was done with both patients with FOG and patients without FOG.

4.4.5 Gait Parameters Estimation

Algorithms for the correct estimation of the gait parameters were included in the sensor embedded software. Some previous activity and the collection of a labelled database was performed in order to implement the most suitable approach.

As part of a series of controlled tests, patients performed a gait test in which their average step length and average step velocity was measured. These values were estimated through the waist-sensor signals and a novel inverted-pendulum model. The sensor location used in REMPARK provides a different kind and shape of signals than those previously obtained in the literature, that commonly are using the Anterior Superior Iliac Spine position for the reported experiments.

Figure 4.10(b) and 4.10(c) show the acceleration signal from lower back (around L4–L5) and left lateral side (near ASIS) of waist, obtained with the REMPARK sensor. It can be seen that the symmetry among left-right steps is lost in signals obtained from the lateral side. Signals from the left leg are more prominent than those from the right leg which impose new restriction on step detection and step length estimation.

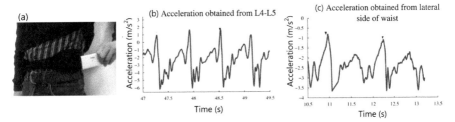

Figure 4.10   (a) The inertial system prototype (9×2, Version 6) positioned in a neoprene belt on left lateral side of waist. Acceleration signals obtained from (a) Lumbosecaral point of waist and (b) left lateral side of waist.

The signals from the lateral side differ from those from the lower back of waist. A newly developed step detection method called SWAT [11] was developed, combined with an adapted step length estimator based to accurately estimate the step lengths from this position. From the left lateral point of view, the proposed gait model considers vertical displacement of waist as an inverted pendulum (IP) model during right step and during single support phase of left step.

Step detection performs an average window that is calculated over the magnitude of the acceleration signals. Mean is removed from this average window signal, and, then, the resulting signal is used to identify left and right initial contacts (IC) and toe-offs (TO) events. When the foot's heal touches the ground, the event is called as IC, and when the foot leaves it is called as toe-off (TO).

The initial contact and toe-off events of left and right legs are noted here as LIC, LTO, RIC and RTO respectively. As the sensor was placed on the left lateral side, the local maximum lateral signal can be used to identify incidents of LICs immediately before or after it. For every local maximum in the SWAT signal, if there was no incident of LIC in the lateral signal immediate before or after it, then it is determined to be a RIC. If there is an incident of LIC, the mid-point from the local maximum to zero is considered to be a LIC. For each detected RIC, the next zero crossing point is considered as a LTO. For each LIC, the mid-point of next zero to the local minimum are searched and considered as a RTO.

The main reference for this work is [11], where it is shown that the results obtained by the proposed method in 28 patients from the REMPARK database show that gait parameters can be estimated with an average RMSE error below 0,04 meters.

4.4.6 Fall Detection Algorithm

The set of algorithms developed and implemented in the REMPARK project is complemented by a fall detection algorithm that was previously developed by one of the partners (UPC). This algorithm enables the detection of falls based on specific computations through accelerometer measurements sampled at 40 Hz. This algorithm is included in the set of algorithms implemented in order to provide more information through the REMPARK system and include the possibility of raising alarms.

The fall detection algorithm has been successfully validated in the "Fall Detection for the Elderly" (FATE) project (CIP-ICT-PSP-2011-5-297178) [12]. It has shown a sensitivity and specificity above 95% along a pilot in which more than 200 users from three countries (Spain, Italy and Ireland) participated.

4.4.7 ON/OFF Motor State Estimation

The algorithmic part for determining the motor state (ON/OFF state) of a person with Parkinson is very complex, because the high degree of subjectivity included in the construction process of a correct model to be used. A main problem is due to the fact that patients, sometimes, are not able to correctly identify their own symptoms and, in some cases, may confuse them with non-motor symptoms. Additionally, when non-motor symptoms are present, it is even more difficult for these persons a correct identification.

This could be a very compromising situation when a machine learning approach is intended to be used, since the most common gold-standard, in these cases, is the patient-diary where the patient should annotate the experimented symptoms every hour, along the day.

In order to be able to implement an objective algorithmic approach to the problem, the related medical literature was reviewed and useful discussions were organized with professionals for determining as much as possible the set of objective conditions characterizing the ON/OFF states. The most widespread definition of the OFF state is to refer to those periods in which low dopaminergic levels occur, in which Bradykinesia is the most correlated symptom. In addition, one can also use the fact that the appearance of Chorea Dyskinesia is commonly produced by high levels of dopamine.

This approach makes possible, based on the algorithms of motor symptoms that have been discussed along this section, to approximate the motor states of the patient with the help of a decision tree technique:

- The algorithm determines that the patient is in ON state when either non-bradykinetic gait or Dyskinesia are detected.
- OFF state is assumed when bradykinetic gait is detected.

This algorithmic approach was tested in the final pilots of REMPARK and its output was compared to the diaries annotated by patients during 3 days. Results are presented in the Chapter 9 and the original public deliverable document (with the reference D9.2) is available at the REMPARK website. The specificity and sensitivity on detecting OFF and ON motor states in 33 PD patients was 89% and 98%, respectively.

4.5 Conclusions

This chapter has presented a huge effort made by REMPARK consortium in order to develop a system capable of monitoring PD motor symptoms in ambulatory conditions. A highly accurate database of labelled signals and clinical questionnaires were collected from 92 PD patients, with more than 340 hours of recorded signals. The labelled signals have been used to train different machine learning methods. The resulting approaches have shown that the selected PD motor symptoms can be accurately monitored through the corresponding sensors, with specificities and sensitivities about 90% in most cases.

Many different algorithms and their results have been presented. The algorithms covering Bradykinesia, Dyskinesia, Tremor, Freezing of Gait (FOG) and gait parameters, employing the collected REMPARK database have been commented and their results presented. However, the ON/OFF algorithm was only tested in the final pilots, since it had to be validated with ON/OFF diaries filled by the patients, used as gold-standard.

References

[1] REMPARK Deliverable D1.2 – Questionnaire addressed to doctors/physiotherapists. Answers and statistical results. Publicly available.

[2] Agid, Y., Ruberg, M., Hirsch, E., Raisman-Vozari, R., Vyas, S., Faucheux, B., Michel, P., Kastner, A., Blanchard, V., Damier., P., Villares, J., & Zhang, P. (1993). Are dopaminergic neurons selectively vulnerable to Parkinson's disease? Advances in Neurology, 60, 148–164.

[3] Braak, H., Del Tredici, K., Rub, U. et al. (2003). Staging of brain pathology related to sporadic Parkinson's disease. Neurobiol Aging 24: 197–211.

[4] Chase, T. N., Juncos, J. L. Fabbrini, G., Mouradian, M. M. (1988). Motor response complication in advanced Parkinson's disease. Function Neurol 3(4): 429–436.

[5] Dauer, W., & Przedborski, S. (2003). Parkinson's disease: Mechanism and models. Neuron, 39, 889–909.

[6] Hughes, A. J., Daniel, S. E., Kilford, L., Lees, A. J. Accuracy of clinical diagnosis of idiopathic Parkinson's disease: a clinico-pathological study of 100 cases. J Neurol Neurosurg Psychiatry 1992; 55: 181–184.

[7] Pérez-López, C., Samà, A., Rodríguez-Martín, D., Moreno-Aróstegui, J. M., Cabestany, J., Bayes, A., . . . & Sweeney, D. (2016). Dopaminergic-induced dyskinesia assessment based on a single belt-worn accelerometer. *Artificial intelligence in medicine*, *67*, 47–56.

[8] Samà, A., Pcrez-Lopez, C., Romagosa, J., Rodriguez-Martin, D., Catala, A., Cabestany, J., . . . & Rodriguez-Molinero, A. (2012, August). Dyskinesia and motor state detection in Parkinson's disease patients with a single movement sensor. In *2012 Annual International Conference of the IEEE Engineering in Medicine and Biology Society* (pp. 1194–1197). IEEE.

[9] Ahlrichs, C., & Samà, A. (2014, May). Is frequency distribution enough to detect tremor in PD patients using a wrist worn accelerometer? In *Proceedings of the 8th International Conference on Pervasive Computing Technologies for Healthcare* (pp. 65–71). ICST (Institute for Computer Sciences, Social-Informatics and Telecommunications Engineering).

[10] Ahlrichs, C., Samà, A., Lawo, M., Cabestany, J., Rodríguez-Martín, D., Pérez-López, C., . . . & Browne, P. (2016). Detecting freezing of gait with a tri-axial accelerometer in Parkinson's disease patients. *Medical & biological engineering & computing*, 54(1), 223–233.

[11] Sayeed, T., Samà, A., Català, A., Rodríguez-Molinero, A., & Cabestany, J. (2015). Adapted step length estimators for patients with Parkinson's disease using a lateral belt worn accelerometer. *Technology and Health Care*, 23(2), 179–194.

[12] FATE project website. www.project-fate.eu.

5

Sensor Sub-System

Carlos Pérez and Daniel Rodriguez-Martin

Universitat Politècnica de Catalunya – UPC, CETpD – Technical Research
Centre for Dependency Care and Autonomous Living,
Vilanova i la Geltrú (Barcelona), Spain

5.1 Introduction

The sensor is an important part of REMPARK since it is in charge of capturing
relevant inertial data from the patient's movement pattern. The location of the
sensor is the left side of the waist and its operation is autonomous and done in
real-time. The sensor, described along the present chapter, has an embedded set
of algorithms, already introduced in precedent chapters, and able to determine
indicators for specific movement disorders related to PD, mainly Dyskinesia,
Freezing of Gait (FOG) and Bradykinesia.

Following the requirements derived for this sub-system (discussed in
Chapter 3), the main design principles, among many others, are:

- the achievement of a device with a reasonable operating autonomy
  (adjusted power consumption), but keeping a small physical size;
- a device capable of being worn comfortably, permitting the regular
  activities along the day; and
- the local execution, in real-time, of the embedded algorithmic set for
  symptoms' detection.

The system includes the classical elements of an Inertial Measurement Unit
(IMU), together with a dedicated part for battery control and energy consumption optimisation. The battery level and the status of the main application
process are indicated to the user using a LED.

A microcontroller (μC) is in the nucleus of the sensor, being responsible for
the real-time and local execution of the developed algorithms on the acquired
data. This microcontroller is also responsible for handling and controlling the
rest of the components.

5.2 Sensor's Data Processing Flow

Before entering into the details of the sensor sub-system, it is necessary to establish the Processing Data flow implemented within the sensor device for executing the symptoms' detection algorithms in real-time.

Developed algorithms operate on acquired samples on inertial sensors, and Figure 5.1 shows a data flow compatible with the previously presented processing specifications. Since many of the implemented algorithms are based on a windowed analysis of the related signals, the figure distinguishes three main parts:

- A first one is a set of calculations performed after acquiring a sample, which mainly consist in the filtering of the signals to condition them, followed by the execution of the fall detection algorithm.
- A second one comprising the calculations performed at the end of a window time, which comprises the main computations for the monitoring of the motor symptoms (detection of Bradykinesia, Dyskinesia and FoG).
- The third part comprises the computations done once per minute, in which the final output of the algorithms is obtained and shared through the mobile phone.

5.3 Hardware Requirements

According to the introductory section, the internal organization of the sensor sub-system used in REMPARK appears in Figure 5.2. This section will introduce the requirements of some critical hardware parts, together with a timing analysis before the specification of the concrete components chosen for the physical implementation.

It must be noted that during the REMPARK database collection phase, the indicated three triaxial sensors (accelerometer, gyroscope and magnetometer) were used to collect a complete set of signals from as many physical magnitudes related to movement as possible. However, in the algorithm development phase, acceleration signals were identified to be enough to monitor motor

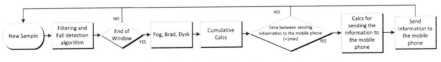

Figure 5.1   Data flow for the algorithms implementation.

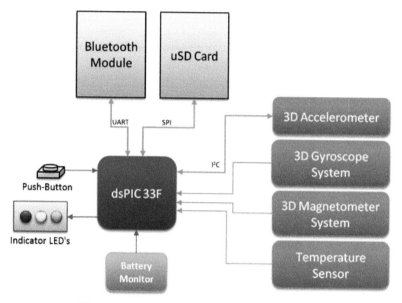

Figure 5.2   Internal organisation of the sensor unit.

symptoms. In this way, it must be indicated that the final sub-system only includes the 3D accelerometer as a sensor.

The most relevant requirements are:

• The concrete component chosen as the correct accelerometer must fulfil the needed range for the acceleration measurements (i.e. full scale values) and to be able to operate at the necessary sampling frequency.
• A main constraint for the concrete microcontroller to be used is the online implementation of the algorithms, so the most important characteristics are memory capacity and the exhibiting computational capacity (throughput).

Memory needs were determined by estimating how many resources should be used by the algorithmic set, in terms of program and data memory. An additional space managed by the firmware, necessary for the storage of the acquired data was also considered.

The theoretical approximation of the required microcontroller computation time is difficult to be done, and for this reason a measurement strategy was organized at the laboratory. Measurements were done around a microprocessor based platform allowing the equivalent simulation of the algorithms.

5.3.1 Memory Requirements

The calculations of the memory size required by the execution of the algorithms and the firmware operation were performed by estimating the online deployment of the algorithms, according the different blocks indicated in Figure 5.1. In concrete, it was considered:

- A combination of algorithms executed on each acquired sample (detection of a fall, conditioning and filtering of the signal).
- A set of algorithms executed per window.
- An estimated need of the window management and the communication system.
- An estimation of the local storage system, using a local μSD card.

Table 5.1 shows the results of these estimations, in terms of necessary minimum size and type of memory (program or data memory).

5.3.2 Sampling Frequency and Full-Scale Values

Sampling frequency and full scale values are key characteristics for choosing the correct accelerometer. Frequency also imposes specific time restrictions on the online implementation of the algorithms, which finally creates important constraints on the specific microcontroller to be used for the implementation.

Table 5.1 Estimated memory usage

Description	Memory Type	Memory Usage (bytes)
Basic System with window Management and communication	Program Memory	6,5 KB
	Data Memory	5 KB
Bradykinesia Algorithm Memory Usage	Program Memory	3,5 KB
	Data Memory	0,5 KB
FoG Algorithm Memory Usage	Program Memory	3,5 KB
	Data Memory	0,5 KB
Dyskinesia Algorithm Memory Usage (FFT included)	Program Memory	3,5 KB
	Data Memory	0,7 KB
Filters+FallDetection Algorithm Memory Usage	Program Memory	1,2 KB
	Data Memory	0,5 KB
μSD for Debug purposes Memory Usage	Program Memory	2,8 KB
	Data Memory	4,2 KB
TOTAL	Program Memory	20 KB
	Data Memory	11 KB

Going along the different developed algorithms, it is necessary to evaluate the necessary sampling frequency to be applied. For this purpose, and also for considering the main frequency characteristics related to human movement the following points are considered:

- In the case of Dyskinesia, it was observed that this symptom increases the power spectra in the lower frequencies up to 4 Hz.
- In the detection of FOG episodes, it was observed that normal walking has a principal frequency around 2 Hz, while FOG episodes are characterised by a principal frequency in the range of 3–8 Hz.
- The developed method for detecting Bradykinesia uses the accelerometer signals associated with the patients' gait. The frequency content of gait is known to be below 20 Hz.

According the Nyquist theorem, the minimum sampling frequency should be 40 Hz. Therefore, the waist sensor must incorporate an accelerometer with a sampling frequency of at least 40 Hz. In order to keep the computing resources as low as possible, this minimum frequency is set as the real-time acquisition frequency.

Moreover, regarding the full-scale range of the sensor, according to the measurements obtained in the signals collected along the project, it is determined that a Full Scale of 6 g (where $1 \text{ g} = 9.8 \text{ m/s}^2$) for the accelerometer is enough for the analysis of human movement with a sensor worn in the waist.

5.3.3 Time Restrictions on the On-Line Implementation

As previously mentioned, the estimate for the processing time requirement was made on a set of measurements of processing time spent by a microprocessor, configured to work at 40 MIPS, and running specific algorithms that have an equal or greater burden compared with the algorithms developed in the project for symptoms' detection. This is an indirect way to fix minimum processing requirements for the microcontroller to be included in the sub-system.

Considering that some operations must be done after acquiring a sample or when a time window has been completed, Figure 5.3 shows the temporal organization of the different computations included in the online implementation of the algorithms, according the data flow presented in Figure 5.1.

The timing of the accelerometer data collection is represented in orange (related to a sampling frequency of 40 Hz, with a time between samples of 25 ms). In the sample zoom (top section of the figure), it is shown the time necessary to both acquire the accelerometer measurements and to perform the algorithm's calculations that are done at every sample.

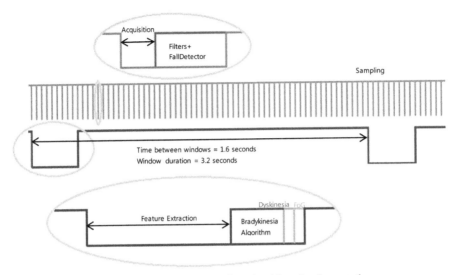

Figure 5.3   Timing for the online algorithms implementation.

The duration of a window, which is 3.2 seconds corresponding to 128 samples, is represented in blue at the bottom section. Note that a new window starts every 64 samples (every half a window), as described in Chapter 4. In the zoomed window, the computations done at the end of an acquired window are shown. These computations start from those provided by the processing done after each sample. Most of the time spent in the window computations correspond to the feature extraction. Once obtained, the bradykinesia, dyskinesia and FoG algorithms are applied. In order to ensure a correct processing, their results are obtained before the end of the next window is reached.

Table 5.2 presents an estimation of the processing time for each part in which the algorithms are divided. This estimation is based on the necessary

Table 5.2   Estimated processing time

Description	Timing	Time (ms)
Sampling Frequency	Between Samples	25
Adquisition	Every Sample	0,014
Filters + Fall detection algorithm	Every Sample	0,0318
Windowing time	Between Windows	1600
Feature Extraction	Every Window (max)	151
Bradykinesia Algorithm	Every Window (max)	55
FoG Algorithm	Every Window (max)	3
Dyskinesia Algorithm	Every Window (max)	3

operations to accomplish each computation in terms of memory usage and the inherent complexity of the calculus.

5.4 Sensor Device Components

This section describes the selection of the components for the sensor sub-system according to the presented requirements.

5.4.1 Microcontroller

The microcontroller (labelled as dsPIC33F in Figure 5.2) is responsible for the management of the data acquisition with a fixed sampling frequency, the analysis of the raw data applying the corresponding online algorithm and sending the processed data (or, alternatively, the creation of a local log in the μSD card).

The selected microcontroller in the project, for these purposes, was the Microchip® dsPIC33FJ64MC804. One of the main advantages of this device is the availability of an integrated DSP engine, which enables advanced computations in short time, when compared with regular microcontrollers (e.g., 32F, 24F, 18F and 16F). The microcontroller memory includes 44KB for program memory while the memory dedicated to RAM reaches 16KB (both are according the requirements indicated in Table 5.1).

This dsPIC architecture is known as "Modified Harvard" which uses 16 bits-long data and 24 bits-long instructions and it processes 40 Mega-instructions per second (MIPS). The DSP engine enhances the operational capacities of the μC. It allows 16-bit data multiplications for both fractional and integers, and it also allows inverse multiplication among 32 and 16 bits of data. The DSP, despite its relative small size, becomes useful when computation power is required, saving time to the main μC threat.

The dsPIC operational voltage range is 3.0 to 3.6 V (the whole system voltage supply is 3.3 V). Working at 40 MIPS and supplied with 3.3 V, the microprocessor consumption is 60 mA in "run mode" (normal operation mode), while in "idle mode" (special waiting state) its power consumption drops to 20 mA, and in energy safe mode ("sleep") maximum current peaks reach 28 μA.

An internal DMA (Direct Memory Access) module allows the communication between the CPU, the memory and the peripherals independently from the process being executed in the main thread of the program, enabling the execution of this process in parallel. The dsPIC has 8 DMA channels that may be associated to any peripheral I/O port. The DMA allows the parallel execution of various processes interrelated among them through asynchronous events (interrupts).

The whole operation of the system is effectively managed by the CPU using some internal specific peripherals, allowing:

- Data acquisition.
- Communication with the wireless module.
- Communication with the digital sensor.

Each peripheral has two associated DMA channels (transmission and reception) that provide access to a shared memory block.

5.4.2 Accelerometer Details

The selected accelerometer is the LIS3LV02DQ (LIS) manufactured by ST Microelectronics [1] which was, at the moment of the design, the only available Microelectromechanical System device (MEMS) using a digital interface within the device. The inclusion of the sensors, the signal conditioning and the converters within the same packaging ensures a very good performance against perturbations. Besides, the inclusion of the three axis (3D) in the same package also provides a better functional and geometrical symmetry. Finally, it is important to highlight that the accelerometer includes a signal compensation based on internal calibration curves, which compensates the measured signal based on the temperature sensor also included in the accelerometer integrated circuit.

The bandwidth used by the LIS is 640 Hz for each axis. Given the particularities of human motion, the sample frequency required for it is 40 Hz, therefore, an excessively small sample period is avoided. One of the main advantages offered by LIS is that the full scale may be chosen between two values (± 2 g or ± 6 g). The higher full scale allows a sensitivity of 340 LSB/g, which is the selected one for the device. The precision of the accelerometer is 2.9 mg for every bit change.

5.4.3 Bluetooth Module

According to the REMPARK system requirements, the movement sensor device has to communicate wirelessly with, at least, a mobile phone to share the output of the algorithms. The system has been developed including a wireless communication Bluetooth v2.1 + EDR (IEEE 802.15.1) chip to implement the communication channel. For this purpose, the sub-system includes a WT12 Bluegiga® communication module, with an on-chip integrated antenna.

The communication is organised through the UART port of the system. When the system boots the connection and communication parameters are configured; once communication is established, and because the SPP profile is used, the module works as a bridge between the UART port and the Bluetooth unit. This module works at 115200 bps and its consumption is 31.5 mA according to its technical specifications [2].

5.4.4 Power Management

The sensor includes a battery management system that tries to save as much energy as possible. As shown in Figure 5.4, the operation is based on 3 low-dropout regulators and each of them powers selected parts of the sensor: first regulator supplies energy to the microcontroller, which manages the operation of the remaining regulators. The additional two regulators supply the analogue and the wireless circuitry.

The system is powered by an 1130 mAh Lithium-Ion battery. According to the functional requirements for REMPARK system, sensor usage is suggested to be similar to a mobile phone. The system has a peak of 100 mA instantaneous current consumption; therefore, the worst case estimated life of the battery is a minimum of 20 hours. Then, the device should be charged a maximum of once per day, and normally during night.

The system incorporates a battery charger and a battery monitor. The battery monitor indicates to the user its current state through a RGB-LED.

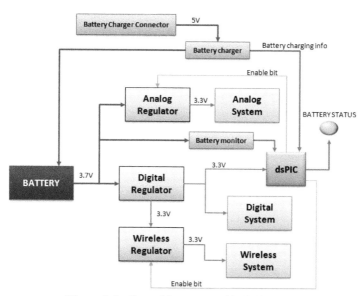

Figure 5.4   Power Management block diagram.

When the battery is very low, an interrupt is produced in the microcontroller and the system automatically closes all its peripherals and communications and enters in the "sleep" mode.

5.4.5 External Memory Unit

The sensor system contains an external memory unit in the form of a μSD card, as it has been introduced, managed through one of the available microcontroller SPI channels. This memory unit is intended to store, if desired, the raw data captured by the sensors. After extracting the μSD card from the sensor unit, it is possible to get the generated log file in order to analyse the inertial data. This functionality was especially useful during the REMPARK database construction process.

5.5 Sensor Casing and Operation

All the described electronic components of the sensor unit plus the Li-ion battery are encapsulated in a 99 × 53 × 19 mm plastic case. The total weight is 125 g (including the battery). The prototype also includes a wall battery charger. Figure 5.5 shows the casing view of the sensor unit.

As parts of the user interface, four elements can be externally identified in the unit: the main switch, an action button, the indicator LED and a charger connector. Figure 5.6 indicates the location of these components.

The behaviour of the sensor unit is determined by two states: Off and On. These states are mutually exclusive; a third state, related to the battery supervision, is compatible with the previous ones and may be understood as

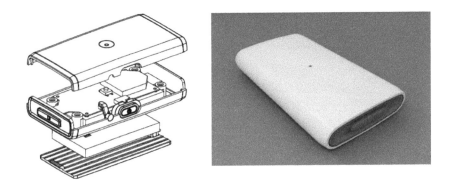

Figure 5.5   Sensor unit casing.

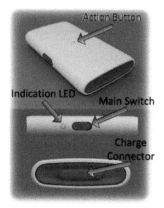

Figure 5.6 External components of the sensor unit.

an independent state. Figure 5.7 represents graphically the relations between the states of the sensor unit.

In the OFF state the sensor cannot work, no battery consumption exists and no battery level may be presented. Besides, the device does not react when the action button is pressed. In the ON state the sensor initializes the microcontroller. When the sensor has already initialized all the internal devices it is ready to start the sensing process. When ready, the sensor reads inertial signals and processes them. Data are sent every minute to the mobile gateway as described above.

The state of the sensor is indicated using a unique multicolour LED by taking advantage of its blinking light. Different states can be distinguished due to the diversity of the colour code of the LED. For example, a blinking green/yellow light means the sensor is in a sense and analysis process. A blinking green/blue light means the sensor is sending data. If the sensor is

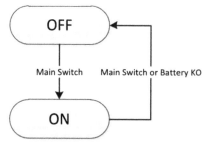

Figure 5.7 Status of the sensor unit.

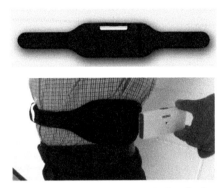

Figure 5.8   Special belt and REMPARK final sensor.

charging the battery, it will be indicated with a fixed orange light. Otherwise, a fixed green light will notify battery if fully charged. Finally, error state will be indicated by means of a fixed red light.

The sensor unit case is surrounded by a retention mechanism (belt) able to fix the sensor unit on the patient's waist, as depicted in Figure 5.8. The part of the belt that is in direct contact with the patient's skin has been manufactured using a biocompatible neoprene material.

5.6 Conclusion

The specification of the concrete requirements has been done along the chapter, allowing the selection of the most convenient components integrating the sensor unit of the REMPARK system. The microcontroller is capable to embed all the developed algorithms and to execute them in real-time and considering all the restrictions related with the data capturing timing.

The system is light and small, being suitable to be worn in a belt without being intrusive for the patient. Furthermore, the system is capable to store raw data and to send messages to an external device. In this way, the developed sensor complies, thus, largely with all the established requirements, becoming the heart of the REMPARK system.

References

[1] STMicroelectronics, Inc. LIS3LV02DQ datasheet. MEMS Inertial Sensor., (2005).
[2] Bluegiga. WT12 data sheet versión 2.5., (2008).

6

User Interaction

Ana Correira de Barros, João Cevada and Ricardo Graça

Fraunhofer Portugal AICOS – Assistive Information and Communication Solutions, FhP-AICOS, Porto, Portugal

6.1 Introduction

One of the components of the REMPARK system is the smartphone, which represents the main communication channel between the system and the user. Within the project, the main goals in the development of the smartphone applications were to complement and confirm information coming from the system's sensors through patient feedback tests and to design user interfaces specifically tailored to the needs and abilities of users with Parkinson's Disease (PD).

Given the nature of a device such as the smartphone and its potential benefits in daily life outside a medical scope, the design and development went beyond the single focus on medical issues to encompass other services, such as the usual communication features offered by a smartphone.

The need for a strong collaboration with end-users was defined at the project onset, which led to an iterative design process. This not only informed choices about patterns of interaction suited for people with Parkinson's Disease (PwP), such as target sizes or suitable gestures, but also about the kind of features main stakeholders would like to see in the REMPARK smartphone. The structure of this chapter will then reflect the two main blocks mentioned above: 1) the research on guidelines regarding visual, interaction and navigation patterns; and 2) the design and validation of the smartphone applications.

The smartphone is increasingly recognized as a viable instrument to be applied in medical diseases with movement disorders [1]. Main applications currently refer to the possibility of monitoring symptom changes over time through the use of relatively economic software. This could potentially

103

represent a significant advance particularly for the management of chronic conditions.

The direct use of smartphones with PD patients should take into account the presence of the motor and non-motor symptoms that could potentially affect the usability of the smartphone applications. It should also be noted that people involved in REMPARK were, mainly, older adults. It was then taken into account that these individuals might not show high familiarity with smartphones, and, more generally, with the use of most modern technologies.

With the exception of a few studies regarding interaction of PwP with desktop solutions or desktop peripherals [2, 3], software [4] or pen-tablet devices [5, 6], at the beginning of project activities, there were few studies on the interaction of PwP with smartphones. There were official guidelines for the sizing of target buttons for people with disabilities and studies on both the sizing and element gap for people with disabilities [7]. Nevertheless, none specifically focused on PwP. This, then, constituted an opportunity for relevant research on and beyond the scope of the REMPARK project, which would likely:

1. help future practice in the development of ICT solutions for this particular target group;
2. help the development of more inclusive guidelines;
3. help contribute to reducing the digital divide through promoting enhanced access to smartphone solutions.

The sub-goals defined for this task were:

- to find suitable patterns for the REMPARK smartphone applications which would enhance accessibility and usability;
- to design the applications themselves under collaborative processes with users so that these applications would be: easily used by people with PD, non-stigmatizing, user-friendly and foster adherence to the medical treatment.

All the user interfaces were successfully designed and the most relevant issues evaluated with end-users. The process also took into account possible extensions to the REMPARK services, which led to specific user interfaces leaving room to future features and additions. The results from the field trials are listed and discussed in this chapter, followed by possibilities for future work.

6.2 State of the Art/Competitive Analysis

As part of the REMPARK activities, a systematic search was conducted in order to capture and analyse relevant information about mobile applications that could assist PwP in managing the disease. The search was organized to find market products and Research & Development (R&D) technology related to this topic, including technology being developed and described in published and publicly available papers and conference proceedings. The search also included currently registered patents to verify the existence of innovative solutions. In terms of products or prototypes, the search was limited to mobile solutions for either Android or iOS smartphones.

Table 6.1 summarizes the results of said search for R&D and market-ready products. The different evaluated parameters are those appearing in the top row of the table.

Some applications, like *Parkinson's Central*, the *Parkinson's Disease by WAGmog* and *Parkinson's Disease Facts* are small applications that provide very useful information about the disease, but do not contemplate any kind of interaction with the user. *PD Warrior* also offers information about the disease and combines it with rehabilitation exercises. *Parkinson's Home Exercises* focuses on the physical side of the disease, providing rehabilitation exercises and cueing support. It does not address the psychological and disease managing sides. *Parkinson's Disease Manager* is an application that provides a relation between the doctor and the patient. It has doctor reminders and questionnaires that allow the doctor to better diagnose the state of the disease continuously.

DAF Professional, *Speech Companion* and *Parkinson's Speech Aid* solely focus on speech therapy, and no other application has approached this method in conjunction with other features. It is an innovative method that is rising among the possible therapies for PD. The *Apple Health Kit* is the only solution we found that can assess PD stage by analysing the voice of the patient. It also covers other features like recording all motor data, questionnaires and, like REMPARK, the Tap games.

Parkinson's Easy Call is the only solution we have found that overhauled the default smartphone applications and adapted them to users with PD. However, it does not include any other extra feature.

Some applications, like *ListenMee, Musical Therapy Service Portland*, *MoveApp* and *Beats Medical Parkinson's Treatment* contain an auditory cueing solution; however the last two also incorporated a few extra features. *Beats Medical*^{TM} also tracks medical intake history and the *MoveApp* is one

Table 6.1 Technology watch on mobile applications for PD management

Application Name	Medication Reminders	Information about PD	Doctor Reminders	Diary	Record Motor Data	Simplified Smartphone Functions	Physical Rehabilitation Exercises	Speech Therapy	Auditory Cueing Solution	Question-naires	Tap Exercises/ Games for Symptoms Assessment	PD stage Assessment by Voice
REMPARK Smartphone Applications	✓ Track intake history	×	✓	Automatic	✓	✓	×	×	✓	✓	✓	×
Parkinson's Central[1]	×	✓	×	×	×	×	×	×	×	×	×	×
Parkinson Home Exercises[2]	×	×	×	×	×	×	✓	×	✓	×	×	×
Parkinson's disease Manager[3]	×	×	✓	✓ Manual	×	×	×	×	×	✓	×	×
DAF Professional[4]	×	×	×	×	×	×	×	✓	×	×	×	×
Parkinson's EasyCall[5]	×	×	×	×	×	✓	×	×	×	×	×	×
Parkinson's Disease by WAGmob[6]	×	✓	×	×	×	×	×	×	×	×	×	×
Parkinson's Disease Facts[7]	×	✓	×	×	×	×	×	×	×	×	×	×
PD Warrior[8]	×	✓	×	✓	✓	×	✓	×	×	×	×	×
ListenMee[9]	×	×	×	×	×	×	×	×	✓	×	×	×
Fox Insight App[10]	✓	✓	×	✓	×	×	×	×	✓	✓	×	×
Music Therapy Service Portland[11]	×	×	×	×	×	×	×	×	✓	×	×	×
MoveApp[12]	✓	✓	×	✓	×	×	✓	×	✓	✓	×	×
Speech Companion[13]	×	×	×	×	×	×	×	✓	×	×	×	×
PD Life[14]	✓ Track intake history	×	×	✓	×	×	×	×	×	✓	×	×
Parkinson Info App[15]	✓	×	×	×	×	×	×	×	×	×	×	×

Apple Health Kit[16]	✗	✗	✗	✗	✗	✗	✗	✗	✗	✗	✓	✓	✓	✓
Beats Medical Parkinsons Treatment[17]	Track intake history	✗	✗	✗	✗	✗	✗	✗	✗	✓	✗	✗	✗	✗
ParkinsonMeter[18]	✗	✗	✗	✗	✗	✗	✗	✗	✗	✗	✗	✗	✗	✗
Parkinson's Diary[19]	✗	✗	Manual	✓	✗	✗	✗	✗	✗	✗	✗	✗	✗	✗
Parkinson Info App[20]	✗	✓	✗	✗	✗	✗	✗	✗	✗	✗	✗	✗	✗	✗
Parkinson's Speech Aid[21]	✗	✗	✗	✗	✗	✓	✗	✗	✗	✗	✗	✗	✗	✗

[1] https://play.google.com/store/apps/details?id=com.apps.parkinsons

[2] https://play.google.com/store/apps/details?id=nl.efox.parkinsonss.en.phone

[3] https://itunes.apple.com/us/app/my-parkinsons-disease-manager/id953530845?mt=8

[4] https://play.google.com/store/apps/details?id=co.speechtools.DAFPro

[5] https://play.google.com/store/apps/details?id=co.uk.org.parkinsons

[6] https://play.google.com/store/apps/details?id=com.quizmine.parkinsondisease

[7] https://play.google.com/store/apps/details?id=com.andromo.dev77961.app379765

[8] https://play.google.com/store/apps/details?id=com.droidhd.pdwarrior

[9] https://play.google.com/store/apps/details?id=com.brainmee.listenmee

[10] https://play.google.com/store/apps/details?id=com.intel.aa.medisense.android

[11] https://play.google.com/store/apps/details?id=com.a1679297990523186 30ed3072a.a17422189a

[12] https://play.google.com/store/apps/details?id=de.speechcare.moveapp

[13] https://play.google.com/store/apps/details?id=com.medando.speechcompanion

[14] https://itunes.apple.com/us/app/pd-life/id430413808

[15] https://itunes.apple.com/us/app/parkinson-info-app/id397358632

[16] https://www.apple.com/ios/whats-new/health/

[17] https://www.appannie.com/apps/ios/app/parkinsonmeter/

[18] https://www.appannie.com/apps/ios/app/parkinsonmeter/

[19] https://www.appannie.com/apps/ios/app/parkinsons-diary/

[20] https://www.appannie.com/apps/ios/app/parkinson-info-app/

[21] https://www.appannie.com/apps/ios/app/parkinsons-speech-aid/

of the most complete solutions found that includes medication reminders, information about PD, a diary, rehabilitation exercises, an auditory cueing solution and questionnaires. There seems to have been less focus on the usability of the system in favour of the rehabilitation methods.

The *Fox Insight App* focuses on the daily management of the disease with medication reminders and a diary. This information is combined with a record of all motor data detected by the smartphone.

In the course of the chapter, it will become clear that, overall, the REM-PARK system offers most of the features represented by all solutions, however it has more focus on the management of the disease, physical and psychological monitoring and improvement of patient Quality of Life (QoL) while using the smartphone applications actively or passively. It does not have any kind of approach regarding rehabilitation or speech related techniques. It also does not provide any kind of general information about the disease. While the lack of these features may be seen as a downside of REMPARK, it is important to state that the REMPARK system was developed as a monitoring system with an Auditory Cueing System (ACS) as actuator that offers assistance in case motor symptoms are detected.

The search in patent databases retrieved the following relevant examples:

- <u>WO2011088307 A2 (Application, filing date 2011)</u> [8]. "A medical monitoring and surveillance system uses a server communicating with a general purpose personal device running an application. The application may be downloadable. The application is configured by the server, i.e. the application configures the device to perform medical tests using the sensors, pre-existing capabilities, and functionality built into the device. The device may be a cellular phone with data communication and other functionality, a personal digital organizer, a portable entertainment device, or another similar personal device. The application reports the results of the medical tests to the server or a third-party device. Various trigger events and associated tasks may be incorporated in the server or in the application residing on the device. A trigger event may occur, for example, in response to the test results meeting one or more predetermined criteria. Once a trigger event occurs, a task associated with the trigger event is performed."
- <u>US8669864 B1 (Grant, filing date 2013)</u>. [9] "A mobile terminal is used to assist individuals with disabilities. A mobile terminal such as a "smartphone" or other commercially available wireless handheld device may be loaded with software. The software may be configured to: (i) store

criteria for managing communications between a disabled user of the mobile terminal and a remote caregiver, (ii) determine whether a criterion is satisfied, and if so (iii) initiate a communication from the mobile terminal to the remote caregiver, and (iv) receive a response from the remote caregiver. Thus, through this software, the mobile terminal may dynamically facilitate communications with specific remote caregivers based on specific situations that may confront disabled individuals."

- US20130345524 A1 (Application, filing date 2013). [10] "Methods and systems are disclosed for sensing and assessing patients' responses to tests using a device that may include tactile input, voice input, still image analysis, and responses to visual and auditory stimuli. In one example, a method includes obtaining interactive clinical assessment data using a remote client device and a computer-based control device, the method including providing on a display of a remote client device one or more test prompts for conducting an interactive clinical assessment, each displayed test prompt instructing a user to perform an action using the remote client device in response to the test prompt, and providing on the display of the remote client device one or more potential responses of actions that may be performed in response to the test prompt."

Analysing the patents found, it is possible to see that none is completely oriented to PwP, however all three describe different methods to manage a system similar to REMPARK. WO2011088307 A2 describes a system similar to REMPARK where a smartphone gathers data from different devices and sends it to the server. After that, some triggers may be configured in both the server and the smartphone. The REMPARK system works differently, as the triggers are preconfigured and the server only calculates the thresholds that activate them. The US8669864 B1 patent describes a simple system of communication between caregivers and patients, also not focusing in PwP. Their focus was in general disabled people, and therefore the applications are adapted accordingly. Given that REMPARK is a monitoring system, the communication between the caregiver and the patient, using the tools developed for REMPARK, is strictly restricted to the management of the disease (medication intakes, medical appointments scheduling, etc.), even though the patient may use REMPARK's smartphone as a normal phone in his/her daily living to establish and receive phone calls, text messages, access the Internet, etc. The last patent (US20130345524 A1) describes the usage of a device's input methods, like touch, voice, gestures, etc. and how they can be used to perform an interactive clinical assessment. The REMPARK system makes use of this method in the Tap games.

6.3 Interaction Guidelines for Users with Parkinson's Disease

The first step towards the design of adequate user interfaces for PwP was understanding which and how specific symptoms of the disease affect the interaction with touchscreen handheld devices such as smartphones; secondly, understanding how said interaction with the smartphone can be improved in order to accommodate the characteristics of the disease [11].

After an intensive literature review on the topic, eight semi-structured interviews with health care professionals who worked with PwP on a daily basis were organized. They were complemented with observation sessions in which two PwP showed their symptoms to their neurologist, as if it was a consultation. The interviews were audio-recorded, coded and analysed, which enabled to understand which symptoms of PD could affect the interaction with the smartphone. Table 6.2 shows some interesting Interview Results (IR) to be considered.

To actually measure the extent to which PD symptoms affect the interaction with the smartphone, usability experiments were created. Thirty-nine participants (17 females, 22 males) took part of the usability experiments. Participants average age was 64 (Median: 66; STD: 7.4) and had been diagnosed as having PD since for at least 10 years (Median: 8; STD: 5.8).

Table 6.2   Some important interview results

Motor Characteristics	
IR1	Bradykinesia can slow repetitive movements.
IR2	Rigidity makes interaction more imprecise and slower.
IR3	Dyskinesia can make the interaction very difficult.
IR4	PD may hinder speech.
IR5	Some PwP may experience visual disabilities.
IR6	PwP are likely to use the phone while standing still or sitting.
IR7	The impact of PD hands' tremor is limited.
Cognitive Characteristics	
IR1	Short-term memory loss is accentuated on PwP.
IR2	Thought is slowed by PD.
IR3	Depression and apathy are common in PD.
IR4	Dementia cases are often observed on later stages of the disease.
General Characteristics	
IR1	Symptoms significantly vary across different PwP.
IR2	Symptoms vary between ON and OFF phases.
IR3	The disease progresses differently from person to person.
IR4	Autonomy is gradually lost.

All participants took part of the experiment while in ON phase. Regarding participant's self-reported motor symptoms: 59% had tremor, 59% had rigidity, and 26% had dyskinesia. Some of them (13%) had undergone deep brain stimulation surgery. The recruitment was done through two delegations of the Parkinson's disease patient association in Porto and Lisbon, as well as the Hospital of São João (Porto) in Portugal.

As it can be seen in Figure 6.1, four different tests were developed to evaluate the following PwP abilities:

1. select targets of different sizes using the Tap gesture
2. do swipes
3. perform repetitive taps
4. accomplish drag gestures

Tap and Swipe were chosen due to their heavy use on today's smartphones. Multiple-tap and Drag were chosen because they were adequate for building smartphone interfaces for medical questionnaires with scales, a requirement of

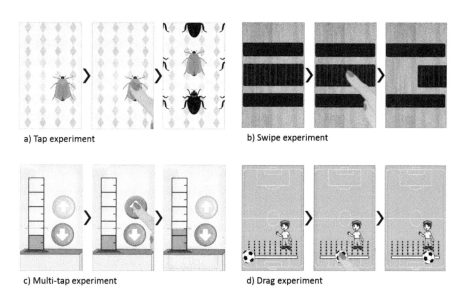

a) Tap experiment   b) Swipe experiment

c) Multi-tap experiment   d) Drag experiment

Figure 6.1 Sequence of interaction of different experiments. a) Participants had to touch the target, which appeared in different sizes, at different positions and surrounded by distractions of different sizes; b) Participants had to slide a rug which appeared in different heights with different spaces to the distractions; c) Participants had to control the water level, by filling the pipette by touching the arrow up until the water reached the green mark; d) Participants had to use the seek bar with a ball as a selector to drag the ball to the boy; scale and mark's position changed with test progression.

the REMPARK project. Experiments were developed for the Android platform and designed to run on the Samsung Google Nexus S (4-inch capacitive touchscreen, with 480×480 pixels of resolution and $123.9 \times 63 \times 10.9$ mm dimensions). All relevant interaction data were logged during the experiments and conveniently analysed.

The results of the usability experiment showed that Tap test results were significant both regarding touch accuracy and reaction time. The results showed that mean touch accuracy tends to decrease with button size, and the highest accuracies were observed with targets with 14 mm size (98% of accuracy). Results also showed that the spacing to surrounding elements does not influence the accuracy of the Tap.

It was observed that PwP are able to perform swipes and distinct participants swipe at very different speeds. Despite target size and spacing between the target and surrounding elements being irrelevant, our results showed that to accommodate 95% of the participants' swipes one should accept movements of 24 mm/s.

PwP were also able to perform successive taps with no significant reduction in speed, which challenged the data from the interviews that anticipated PwP would be strongly affected by bradykinesia. However, as participants performed the tests while in ON phase, this might not be the case.

Regarding the Drag gestures, it was shown that participants were able to drag objects with precision, but they were slow to reach the goal. However, the increased frustration expressed by the participants with the Drag, allows us to say that Multiple-tap is more comfortable to perform than Drag.

By reflecting on the findings of this short study, 12 user interface design guidelines (DG) for creating smartphone applications for PwP were considered and are listed in Table 6.3.

Table 6.3   User interface design guidelines

Design Guidelines	
DG1	Use tap targets with 14 mm of side.
DG2	Use the swipe gesture, preferably without activation speed.
DG3	Employ controls that use multiple-taps.
DG4	Use drag gesture with parsimony.
DG5	Prefer multiple-tap over drag.
DG6	Adapt interfaces to the momentary characteristics of the user.
DG7	Use high contrast coloured elements.
DG8	Select the information to display carefully.

Table 6.3 Continued

DG9	Provide clear information of current location at all times.
DG10	Avoid time dependent controls.
DG11	Prefer multi-modality over a single interaction medium.
DG12	Consider smartphone design guidelines for older adults.

6.4 User Research and User-Centred Design Processes

The choice was to provide the users of the REMPARK system not only with the REMPARK applications, but also common functionalities of mobile phones. The REMPARK applications could have been designed as stand-alone applications which could be downloaded and be amongst the smartphone's existing applications. Two performance-based reasons discouraged this approach:

1. The REMPARK system is a medical one containing sensors and actuators which play a serious role in users' lives. Therefore, the system should have some control over its performance, which means having, amongst others, control over battery duration or some means of preventing external applications' interference with REMPARK's applications.
2. As described in the previous sections, users with PD are generally older adults who, nowadays, are likely to be less tech-savvy and who, additionally, due to the disease itself, have motor and non-motor constraints which might hinder the interaction with the smartphone. These users then require simple user interfaces which are furthermore adapted to their abilities.

These reasons have led the partners to develop a set of applications which would work within a launcher. In a nutshell, launchers enable developers to create a custom home screen through which one also has the possibility to control what the user will or will not see and interact with. The advantage of the launcher is that it answers the above listed reasons in the sense that it allows to control which applications the user has access to and, furthermore, in what usability and user experience are concerned, it allows for the design of a single interaction experience. All applications within the launcher obey the same principles and patterns and, once the user has learned how to use one, he or she will most likely be able to use the others, as the language is kept identical.

The processes followed a human-centred approach in that main stakeholders were called to be part of the design process from onset. Following the above described reasoning, the initial list of applications to design and implement within REMPARK was set amongst the partners in the consortium as follows:

- Messaging
- Emergency call
- Calling
- Contacts
- Show disease status
- Medication
- Agenda
- Auditory Cueing System (ACS) controller
- Medical questionnaires
- My data
- Settings
- Tutorials

After this definition, a process of interaction with users was scheduled in order to cope with main usability requirements, not previously considered.

Apart from the health-related benefits arising from the ACS, the value proposition for this set of REMPARK applications would be that PwP could have all their Parkinson's-related mobile needs satisfied, while not hindering the use of the main purpose of a smartphone: communication. All aspects of the applications – ranging from button sizes to workflows – were taken into account from users' perspective in order to insure a positive user experience.

After informal discussions with the experts within the REMPARK project, who provided insights about PwP's needs and abilities, app design was approached first through an exploratory methodology, in semi-structured interviews to people with Parkinson's disease, informal caregivers and physical therapists. The analysis of these results informed a second phase for which a collaborative and participatory approach was considered, mainly involving people with Parkinson's disease and their caregivers.

The results were not conclusive regarding most issues (medication/appointment reminders, preferred activities or physical therapy). On the one hand the samples were fairly small and, on the other hand, participants with PD gave ambiguous answers to the questions posed by the researcher. Nevertheless, there was a clear concern with moving around outside (especially in places further away from home) and with having emergency mechanisms at PwP's disposal in case something happens.

These interviews helped to structure the next steps in user research and to add a new requirement for the smartphone: that it should provide others the information that the smartphone owner has PD.

The material up until this point had enabled the design of the user interfaces and some screen flows. The choices for the UIs and, specifically, the GUIs were made based on Fraunhofer Portugal's existing knowledge and expertise on UIs for older adults, topped with the findings regarding specific requirements for users with PD.

The background was kept dark to prevent glare and increase battery life and all buttons size complied with the recommendations mentioned in Table 6.3. The approach was to avoid the need for scrolling actions to the widest extent possible. Therefore, activities were sometimes divided into layers and, consequently, sub-activities. All screens consistently displayed the same Back button, appearing at all times and in the same location. All action buttons were consistent in displaying, whenever possible, an icon along with an action verb + name, so that the user would be surer of his or her input. For instance, whenever possible, instead of "Save", the button would read action verb plus the element being saved, e.g. "Save Appointment".

Regarding navigation, the screen flows were kept with as few layers as possible, wording was sought to adapt to terms which were familiar to users and which could have relation to the physical world. For the composition of the screens we have opted to design a set of patterns for a number of situations/functions, which allowed to be grouped according to each situation. Figure 6.2 shows some examples of the designed patterns.

In order to allow for as simple as possible an interaction, the text input created a new activity so as to allow for a simple screen which would not demand too much of the users' attention. As seen in Figure 6.3, once activating a textbox, the user is led to a screen with a title reminding the input action, a textbox, a save button and the keyboard. In this way, the user was offered step-by-step guidance on information input.

Scoping sessions with stakeholders provided additional ideas for features which would enhance the REMPARK smartphone, such as the different ways to trigger an emergency alert, the importance of showing information about the disease, along with some clues on what to show, the importance of having clear information on medication reminders in order to reduce error, or the usefulness of keeping a detailed record which both the medical doctor and the PwP/caregivers can consult.

With all the compiled information and generated inputs, the partners proceeded to redefine parameters in the REMPARK applications. At this time, the screen flows for the following applications had been re-defined:

• Medication
• appointments

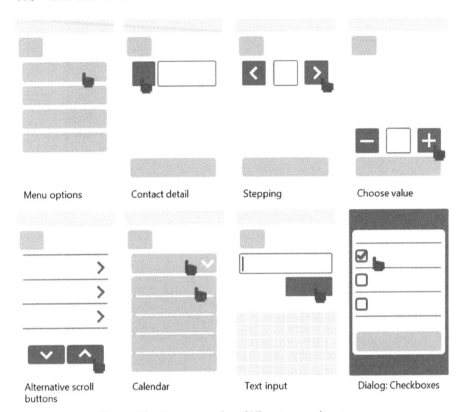

Menu options   Contact detail   Stepping   Choose value

Alternative scroll   Calendar   Text input   Dialog: Checkboxes
buttons

Figure 6.2   Some examples of UI patterns and gestures.

Figure 6.3   Sequence for text input.

- calls
- messages
- contacts
- 'my data'.

A specific interaction session with PwP was organized at this point with the objective of clarifying some issues within the medication application, namely the amount and type of information being displayed (in *medication*, *disease status*, *appointments* and *my data* applications) and to check whether or not the flow was easily followed. The session also aimed to discuss the options for emergency calls with the participants, the option of having SMS templates and the option for the appointments application to have a rough prediction of OFF states. Finally, participants were asked about what they thought would be the best icon for the Auditory Cueing System controller. Some comments on this session:

- **About the medication.** Participants were asked to state their opinions regarding the proposed solutions and also to contribute with their critical thinking and experience with the disease. For the most part, the solutions were welcomed by participants with minor remarks. For instance, regarding the 'Medication Details', participants thought there would be no point in displaying the name of the active substance along with the medication name. Participants agreed that the possibility to add notes to the medication details was useful and that these should show on the reminders (e.g. 'Sinemet 10/100 mg, Take before eating').
- **About emergency situations.** Facilitators discussed the ideas of having one lateral hardware button reconfigured to act as quick dial to an emergency number on long press. They also discussed the quick dial to a caregiver and the permanent presence of a quick dial to an emergency number through an icon on the smartphone's main screens as it is shown in Figure 6.4. Participants tried out long presses to the lateral hardware buttons and agreed that it would be an easy action for them to complete. They envisioned that a possible use case would be an emergency during bathing.
- **On "My data".** Participants were also happy about the possibility to have a quick access to their health information under 'My Data'. At this point in time, the 'My Data' flow would first show a pop-up with a warning to inform that the smartphone owner had PD, followed by a screen with the owner's picture, a quick dial button to the caregiver and a list of current medication. Participants mostly cherished the possibility to carry with them the list of medication they were under at any point in time.

Figure 6.4   Screenshots of the home screens: Applications screen (to the left) and Favourites screen (to the right).

After the advice of medical partners, the adopted solution is presented in Figure 6.5, where all relevant information is shown at once (preferable avoiding scroll).

- **On the Disease Status.** Participants welcomed the idea of visualizing their disease status. They would appreciate it if the application would be able to record involuntary movements (dyskinesia). One caregiver in the

Figure 6.5   Screen flow for the 'My Data' application.

group had already her paper version of this disease status visualization, which she would take with her every time she accompanied her sister (PwP) to the doctor.

According with medical partners, one of the main drivers for choosing to include such an application was to promote treatment compliance. Several studies show the nefarious consequences brought about by non-compliance with medical treatments [12, 13]; nevertheless, controlling medical compliance without being intrusive and tiresome to users is a difficult task. Our hypothesis was that, by allowing users to easily visualize the relation between their disease status and their medical treatment adherence, they would be compelled to follow the medical prescriptions and advice. However, medical doctors pointed out that, if a PwP is not doing well, visualizing a poor health state might have the contrary effect.

Therefore, the choice was to present simple data gathered by the system, which PwP usually try to collect in order to show to their physician. The features to be shown in this application were: ON/OFF periods, FOG, dyskinesia, festination and falls. Along with this, the visualization tool would also show *failed medication intakes*, rather than all data about medication. The final result may be seen in Figure 6.6.

Figure 6.6 Screenshot of the 'My Day' app.

- **On the Appointments.** The idea brought forth about warning the user whenever he or she was setting up an appointment which most likely overlapped an OFF phase was not popular. After some initial discussion, participants agreed this might be dangerous not only because – given the several variables involved – it is quite difficult to predict an OFF timeframe, but also because having this information beforehand could most likely create unnecessary anxiety.
- **On the ACS icon.** Participants decided the best icon to convey the idea of auditory cueing between some elaborated and proposed ideas.

Despite the effort placed onto adopting inclusive design for UI design, the users' concerns about the response of the touch screen led to conduct more research into the touch patterns of PwP in order to try to define what characterizes an intentional tap, and thus optimize interaction with the smartphone by reducing error. For the purpose, a specific test was designed and carried out with 4 participants with PD.

The tests were designed to explore the possibility of creating an algorithm to differentiate voluntary from involuntary taps using the touch data provided by the Android API. For that a little game was designed to persuade PwP to do voluntary taps on the smartphone's screen, consisting in basic math operations where the user had to answer a question by tapping one of 3 possible answers (Figure 6.7a). This game allowed to easily observe when a person would intentionally tap an option or tap on another part of the screen by chance.

It was planned to achieve at least 11 voluntary taps per participant, in this way 11 questions were presented and if they failed solving the problem at

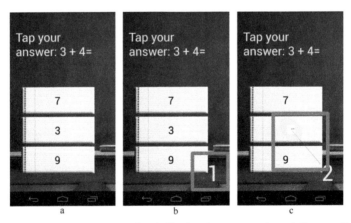

Figure 6.7   a) Initial screen; b) Number indicating the tap counting; c) Custom display of a tap event.

the first time they would need at least one more tap to get the next question. However, we could not induce any involuntary taps from the participants; therefore, no minimum number of involuntary taps per participant was set. The tests were recorded on video and each tap would increment a number on the screen (Figure 6.7b), so after the test was done it was possible to discriminate if a tap was voluntary or not.

The data provided by the Android API consisted of tap time, area of touch, pressure of touch and dislocation of the finger during the tap event which could be used to both perceive the orientation of the finger. The area of touch and pressure were only estimated by the Android API by measuring the change in capacitance on the screen, therefore not corresponding to real measures. As a result, a visual representation on the screen of these measures was created, as a means to test their correspondence to real measures. The pressure was represented by an ellipse that we then stretched and rotated accordingly the perceived orientation of the tap event, the area by a square and the orientation by a red line (Figure 6.7c).

Despite inconclusive, the results suggested that there should be room for improvement and that this analysis should be pursued further in the future. An interesting correlation between the contact time in voluntary and involuntary taps was found. However, the pressure and area had little fluctuations; the direction of tap also did not prove effective because PwP lacked a consistent movement that was necessary to detect correctly the orientation of the tap gesture. But nevertheless, the information about the dislocation of the finger could be used to mark involuntary taps if the finger moves too far away from the first point of touch. Hence, the most probable outcome of future tests will be a time-distance frame where a voluntary tap is considered as a tap that respects that time-distance frame, and involuntary tap as one that does not respect that time frame. This timeframe would be necessarily adjusted by the user in order to better suit his/her needs.

Following these very concerns in optimizing interaction and preventing errors, the velocity of scroll was adapted in order to better control for the presence of dyskinesia. Furthermore, it was added the option to replace the scroll gesture by tap buttons with upward and downward (Figure 6.8).

6.5 Final Usability Tests

For the purpose of conducting comprehensive usability tests with PwP, partic-ipants were recruited through Hospital de São João, in Oporto (Portugal). The documentation was sent to the Ethics Committee and approval was obtained in due time upon documentation submission.

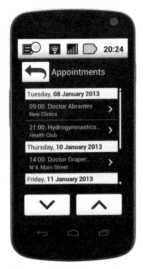

Figure 6.8   Arrow buttons as an alternative to the scroll gesture.

6.5.1 Research Questions

The main questions left unanswered were related to touch interaction, ease of navigation and the affordances of particular patterns. Regarding navigation, most questions were brought about by the two most complex applications: Medication and Appointments. These, it was supposed, will be frequently used and imply an array of activities and tasks which might be difficult to handle, such as interacting with reminders, introducing new data, browsing and searching for specific data, amongst others. Therefore, usability tests focused on evaluating the ease of understanding of the navigation experience, while, assessing the touch and gesture interactions on the background.

6.5.2 Sample

The sample was recruited from the neurology department, a group of a physiotherapist's patients and a day care centre. No exclusion criteria were set, except for the presence of concomitant health conditions which caused motor or cognitive impairments beyond those owing to PD. A total of 12 participants were recruited, although three were excluded for not being able to finish the usability tests due to poor mental state. Consequently, the sample was reduced to nine participants, which resulted on five tests for Appointments, five tests for Medication and four tests for scroll accessibility. The sample had an average

age of 71.3 years old, ±STD 10.4, (median: 73) and participants had had PD for an average of 9 ±STD 8.6 years (median: 6).

6.5.3 Protocol, Script and Metrics

As there were many items in need of evaluation, there was the risk that usability sessions would be too lengthy and hard on participants. Therefore, the session was devised into two different session tests, one for each application: Medication and Appointments. In this way, beyond avoiding wearing out the participants, bias was also avoided to some extent, as participants would not learn visual and interaction patterns from one application to the next.

The protocol followed the usual steps in usability testing: researcher presentation, project presentation, material presentation, study presentation and goals, obtaining informed consent, session tasks and debrief.

The variables being measured were as follows: tap button precision, tap length time (evaluated with FUSAMI [14]), task completion, number of errors and scroll errors (measured afterwards through video observation and with the specific test). Additionally, in order to continue to adapt the sensibility of the screen to better fit PwP the touch data from the appointment tests was catalogued in voluntary taps and involuntary taps.

The scroll test was done at the end of the 'Appointments' session and it had a simple design, it consisted of a contact list with thrice the number of contacts supported by a screen with no need for scroll and the person had to tap on a contact which was only accessible through scrolling. For the 3 types of scroll all the names of the contacts changed, but in all of them the contact was in the last position of the second scrolling panel, thus being required to do an entire scroll to see it; scrolling all the way to the end of the list would hide the referred contact once more. The measures used were: number of touches on the screen and time to find the contact.

The test material, protocol and script were subjected to pilot testing with two participants with PD for each test, who were recruited through Fraunhofer's contacts with day-care centres and who signed the informed consents to take part in the pilot tests. After the pilot tests the session plan was refined.

6.5.4 Results

Some results obtained on the main issues addressed during the set of usability tests are discussed:

6.5.4.1 Appointment test

Using the calendar and reading an appointments list. Overall the participants recognized which days contained appointments (Figure 6.9), even though some had more difficulties than others. In the end, all participants identified the days correctly. Current day is also visually distinct from the remaining month days, but only one participant noticed this.

As expected, when asked to see more information about an appointment they had on a specific day, participants tried to press on that day, despite being too small for participants to accurately tap it. Participants were able to read the correct event. They only had difficulties when the event did not show immediately on the list and they had to scroll to get the correct appointment. When the facilitator scrolled to the right position they read it without effort.

Tapping on an appointment to see more details. In general, the participants were able to correctly identify the name of the appointments as a clickable element and tapped them without almost any problem (Figure 6.9).

Editing/creating an appointment. Here the participants had more difficulties to know what they had to do in order to edit an appointment (Figure 6.10). Creating a new appointment was the most complex task on the tests and the participants were only able to accomplish it with assistance. Generally, the editable elements were easy to identify, however when they were 'hidden' by the scroll, participants needed help to find them. Additionally, the pop-ups confused some participants. When the pop-ups of date and hour

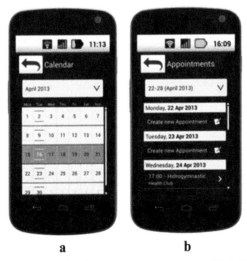

a   b

Figure 6.9   Calendar screens: a) Calendar; b) Week list.

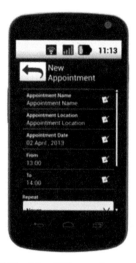

Figure 6.10   New/Edit appointment screen.

appeared on the screen some participants instinctively tried to tap on 'Set' before using the controls on the pop-up to set a different hour/date.

Using scroll. No participant, with the exception of the one who was used to touch technology, had the initiative to use scroll. After its use was shown to them, participants were able to replicate it in that situation but rarely used it afterwards. Even though they were able to do the gesture, they would only perceive the information currently displayed on the screen and would not seek more information. Participants did not notice the visual marking on the screen indicating the scroll.

6.5.4.2 Medication test

Reading the medication list. Almost all participants correctly identified the medication names and notes with almost no effort (see Figure 6.11).

Tapping on a medication to see more details. When asked to see more details, the majority of the participants correctly tapped on the medication to disclose more details about it, but they had difficulty in correctly reading the intake hours for that particular medication. As the intakes were on a different list from that of other elements, only after reading all the details regarding that medication (or getting some hints from the facilitator to seek more information), did they finally read the intakes; again, with difficulties.

Edit intake hour and frequency. After understating the concept of intake, it was easy for the participants to tap over the intake event to edit it. Concerning

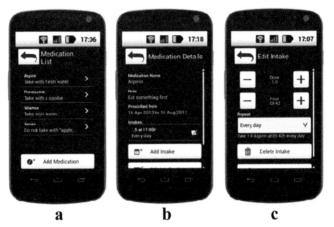

Figure 6.11   Medication apps screens: a) medication list; b) medication details; c) edit intake.

the editing mechanisms, and despite some initial confusion, participants were able to use them correctly. After some taps they understood what the '+' and '–' controllers were for and used them. Participants were also able to recognize the spinner to change the frequency of the intake, and choose the correct days although with hesitation on what the options: "Custom interval" and "Custom week days" would do. Again, they were not able to scroll down to tap the 'Save' button.

Read the medication schedule and list of missed intakes. Despite the majority of the participants having correctly identified the hours and medication which they had to take, some had difficulties associating the hours appearing on the list with the correct medicament. Generally, participants that understood the concept of missed intakes were able to say which medicaments were not taken.

Interact with the medication alerts. Overall, the participants understood clearly which medicaments they had to take, but some had difficulty understanding what they had to do in order to inform the system that they only had taken one medicament of the displaced list (Figure 6.12). However, with the help of the facilitator showing them how tapping on a medicament would mark it with a check icon, they would then proceed to tap over 'Taken' to finish the task.

Voluntary and involuntary taps differences. The test allowed us to observe differences in touching times between some involuntary and voluntary taps (Table 6.4), though many times those differences were not present, making it not possible to differentiate them in an automatic way. However, it is possible

Figure 6.12   Notification pop-up screen (medication reminder).

Table 6.4   Mean duration of a single tap for each participant

	Involuntary		Voluntary	
Participant	Mean of Time (ms)	Std. Deviation of Time (ms)	Mean of Time (ms)	Std. Deviation of Time (ms)
1	166.62	197.38	133.19	90.64
2	355.00	295.57	373.06	250.11
3	322.78	481.36	241.67	217.59
4	413.53	456.89	193.50	87.46
5	703.25	281.18	292.92	178.22

to do it if the PwP intentionally changes his/her behaviour while touching the screen. Since most of the time the involuntary taps are quick (when one has tremor and just passes the finger over the screen, or low dexterity and notices that he/she is touching intentionally on the screen and moves the finger right away), a minimum time threshold could be settled to discard those taps and PwP would always have to press the button above that threshold to have a valid tap. Additionally, a maximum time threshold could be also settled for the case that the person is touching on the screen and does not notice it, thus also discarding this possibility of involuntary tapping.

6.6 Conclusions after the User-Centred Process

This section presents a list of some generic conclusions after the user-centred design approach process organized in the frame of REMPARK, giving, in this way, the opportunity to generate a good solution for the interface implemented on the used Smartphone along the pilots:

- **Concerning the Screen Interaction:**
  - There were no noticeable difficulties on the accuracy of the tapping action when buttons size was reduced (elements measuring 10,5 mm in height were tested).
  - Concerning the test on the best approach to scrollable screens speed (normal or reduced), participants were able to use both. There were no enough data to conclude something relevant.
  - When participants were using button arrows as an alternative to scroll gestures, they were able to learn and subsequently use the arrows approach. In principle, it was noticed that it was not intuitive for them.
  - It is not possible to discriminate voluntary taps from involuntary ones although it might be possible to induce new tap behaviour to PwP in order to make that distinction possible.
  - In general, the users had no problems using spinners either on the Medication or the Appointments applications under test.
  - Concerning the action produced when clicking on "+ and –" buttons, some users had difficulties using them at first, but after some time or after getting some hints, they got used to them and were able to finish the tasks using these.
  - Participants were able to use the pop-up dialogues to change date and time. Some were confused at first and used the down arrows to increment the values instead of the up arrows; but then quickly learned from the mistake.
  - No user had problems tapping over the checkboxes.

- **Concerning the Medication:**
  - In general, the participants were able to read the medication list and details. The main difficulty was in reading the intakes list for a given medicament.
  - The most part of participants had no problems reading the medication that they supposedly had to take for the current day.
  - After some hesitation on the options 'Custom interval' and 'Custom week days', users were able to understand the intake repetition options.
  - Most participants were able to identify the medication that they had missed.

- After some interaction with the researchers, users were able to understand that they can modify a missed intake by inputting the real intake hour.

- **Concerning the Appointments:**

  - Participants were able to read correctly which days had appointments, although they missed the current week and the current day markings.
  - It was difficult to understand the tree view of Month > Week > Day. Nevertheless, they were able to use the calendar and read the appointments for a specific day.
  - Overall participants were able to correctly read the date of the events presented on a list, although there were some exceptions.
  - Users were able to understand that an event can be repeated multiple times on the calendar by editing it.
  - When users tried to add a new appointment, they needed help to type on the phone, and scroll down to see more options and the save button.
  - In general, it was difficult to use and to add alarms to an event.

6.7 Characteristics of REMPARK Smartphone Applications

Considering all the previous presented user-centred process and the preliminary conclusions obtained, a final implementation of the smartphone applications and user interface was organized according with the following main ideas:

- **On Home Screen**
  In order to avoid scroll and allow users to have a quick overview of each block of applications, the home screen was divided into three main blocks: General applications, Favourites and REMPARK applications.
- **On Medication**
  The Medication application retrieves and displays the neurologist prescriptions. At the same time it allows the user to add other medication which is not Parkinson's-related. Users can edit this latter medication, but not the one which has been placed into the system by the medical doctor.

The application shows the list of current medications the user is on, missed intakes, a schedule of intakes and a record of intakes.

The application shows medication reminders at intake time and the user may choose to keep track of his intakes. In this case, the medication reminder displays an option reading 'Taken'. The option for the wording was to prompt users to press the option 'Taken' only after they have taken the medication. This was done to provide users with a mechanism to help them know whether or not they have taken the medication. If, by some chance, the user has taken the medication but forgot to record the event on the smartphone, he or she can go to the 'Missed Intakes' list and edit it by choosing that particular missed intake and editing it by inserting the hour at which the medication was indeed taken.

- **On Appointments**
As with the Medication application, this application retrieves the information from the REMPARK server on existing appointments made by the doctor, while allowing the user to add their own. Again, users are only allowed to edit the information which has been inputted through the smartphone itself.

The Appointments application allows the users to access a calendar view (and reach their appointments from there), shows a list of events ordered by date and allows the user to create new events from the main menu.

- **On Auditory Cueing System Controller**
The auditory cueing system controller application has a single screen and allows the following basic options: turn on/off, change volume and change rhythm. Users may also choose to change the sound type by previewing and selecting it and predefine the settings for the auditory cueing. This is done through the settings application.

- **On Medical Questionnaires**
Medical questionnaires pop-up as they are fed into the system. Nevertheless, the user may opt to dismiss the questionnaire and answer it later. In this case, the Questionnaires application icon displays a notification showing the number of questionnaires waiting to be completed.

Due to the limited available space in a single screen (i.e. without scroll), there will be limits imposed to the number of characters for questions and answers. The reasoning behind this choice lays in the fact that the smartphone is one of the possible ways to answer medical questionnaires. The use cases predict that more complex or longer questionnaires may be answered through the web interface.

- **On My Data**

  The My Data application is meant for emergencies or for a quick overview of the user's health data to display, for instance, in a visit to the doctor who is not part of the REMPARK system. After a pop-up with a warning about the person's disease (meant for emergency scenarios), it shows the smartphone owner's picture, his or her name and phone number, along with health information: blood type, medication list, concomitant illnesses and specific data about Parkinson's disease.

- **On My Day**

  The My Day application refers to what was at first called 'Disease Status' application. The name was changed to a more user friendly working. It has a single screen displaying information of the user's state over the previous day: ON/OFF periods, FOGs, festination episodes, missed medication intakes, falls and dyskinesia.

- **On Quick Emergency Call**

  This application works as a quick dial to an emergency number and is displayed both at the Applications and Favourites screens.

- **On Calls**

  The Calls application allows the user to make a phone call through a list of contacts or by typing in the number using an extra-large dial pad. It also allows checking missed calls and call history.

- **On Contacts**

  This application displays a regular contacts list and allows the user to define favourite contacts, call or send a message to a given contact, add new contacts and delete existing contacts.

- **On Messages**

  This application allows the users to simply read, listen to and send messages (through text or speech). It also supports sending and receiving pictures.

- **On Settings**

  The Settings application allows the user to set his or her preferences regarding Sounds, Vibration, Wi-Fi, Accessibility, Language, Auditory Cueing Settings, to set a Caregiver (which will show at the Favourites home screen and to Add a new device. In order to allow for troubleshooting, this menu shows an option which allows users to check the status of the sensors and actuators in the REMPARK system.

- **On Status Bar, Notifications and Shortcuts**

  The status bar is shown at the top of all screens and it displays information about the time, the battery status, the Wi-Fi connection status and strength of cellular signal. Furthermore, the status bar is used to host shortcuts for cueing controllers and to provide users with straightforward information about their status. When ACS cueing icons is on display in the status bar, they act as buttons to the applications. The notifications pertaining to each application will be shown on the home screens, on top of the corresponding application icon.

6.8 Some Concluding Remarks

The presented interface was completely integrated in the smartphone and used as part of the whole REMPARK system during the complete trials organized as one of the main activities at the end of the project. Participants were given the smartphone along with a user manual and, in the end, were asked to fill in SUS (System Usability Scale) and QUEST 2.0 (Quebec User Evaluation of Satisfaction with assistive Technology) questionnaires. Therefore, participants evaluated the whole system, not just the smartphone user interfaces.

The results indicate that there are some issues in terms of learnability, which should be looked at in the future. As it was reported during evaluation of REMPARK, "...*managing all that REMPARK has to offer at once can be quite overwhelming, especially if one is using the system for a short period of time and if the user's level of familiarity with technology is not very high. REMPARK is a completely new and innovative system and it is only natural that these barriers emerge. In the future, however, we could try to work on this part of introducing the system to new users, covering the unboxing experience as well. But this is for the future, as it was never part of the goals for our project*".

It should also be noted that each participant interacted with the system for four days, which could explain why, at the end of the trials, participants were not completely familiarised with all the REMPARK apps. Nevertheless, satisfaction levels, as reported in the scales used, were quite high – for the participants, the REMPARK system was comfortable, secure and safe.

Finally, one of the most interesting found discoveries during the field trials had to do with the 'My Day' application. The app was not planned at first and was born from listening to PwP's conversations about their symptoms. During the field trials, screen interaction was analysed. The 'My Day' screen showed a high activity with taps on the date and on the elements related to symptoms. This seems to suggest that participants wanted to interact with their own data.

On REMPARK's final workshop, we had the opportunity of showing the apps to Sara Riggare, a PD advocate and PwP herself. We commented on these results and findings, which did not surprise her. However, she recommended to focus on displaying the positive events, rather than the negative ones. Further work on the 'My Day' app following the field trials and the experts' advice was considered by us to be relevant work for the future

References

[1] N. Kostikis, D. Hristu-Varsakelis, M. Arnaoutoglou, C. Kotsavasiloglou e S. Baloyiannis, "Towards remote evaluation of movement disorders via smartphones," em *IEEE Eng Med Biol Soc*, 2011.
[2] L. M. Cunningham, D. D. Finlay, G. Moore e D. Craig, "A review of assistive technologies for people with Parkinson's disease," *Technology and Health Care*, vol. 17, no. 3, pp. 269–279, 2009.
[3] S. Keates e S. Trewin, "Effect of age and Parkinson's disease on cursor positioning using a mouse," em *ASSETS 2005*, 2005.
[4] J. L. Levine e M. A. Shappert, "A mouse adapter for people with hand tremor," *IBM Systems Journal*, vol. 44, no. 3, pp. 621–628, 2005.
[5] P. Maziewski, P. Suchomski, B. Kostek e A. Czyzewski, "An intuitive graphical user interface for the Parkinson's disease patients," em *4th International IEEE/EMBS Conference on Neural Engineering*, 2009.
[6] B. Göransson, "The re-design of a PDA-based system for supporting people with Parkinson's disease," em *People and Computers XVIII – Design for Life*, Lodon, Springer, 2005, pp. 181–196.
[7] K. B. Chen, A. B. Savage, A. O. Chourasia, D. A. Wiegmann e M. E. Sesto, "Touch screen performance by individuals with and without motor control disabilities," *Applied Ergonomics*, vol. 44, pp. 297–302, 2013.
[8] B. Gutekunst, "System and method for medical surveillance through personal communication device". United States of America Patente WO2011088307 A2, 14 January 2011.
[9] D. E. Tedesco, R. C. Tedesco e J. A. Jorasch, "Methods for remote assistance of disabled persons". United States of America Patente US8669864 B1, 11 March 2014.
[10] B. C. Meyer e J. Zivin, "Device and methods for mobile monitoring and assessment of clinical function through sensors and interactive patient responses". United States of America Patente US20130345524 A1, 16 December 2013.

[11] Nunes, F., Silva, P. A., Cevada, J., Correia de Barros, A., Teixeira, L. (2014). "User interface design guidelines for smartphone applications for people with Parkinson's disease". Universal Access in the Information Society, 16(2).

[12] L. R. Martin, S. L. Williams, K. B. Haskard e M. R. DiMatteo, "The challenge of patient adherence," *Therapeutics and Clinical Risk Management,* vol. I, no. 3, pp. 189–199, 2005.

[13] M. Viswanathan, C. E. Golin, C. D. Jones, M. Ashok, S. Blalock, R. Wines, C. M., E. J. L. Coker-Schwimmer, D. L. Rosen, P. L. Sista e N. thleen, "Interventions to improve adherence to self-administered medications for chronic diseases in the United States," *Annals of Internal Medicine,* vol. 157, pp. 785–795, 2012.

[14] FUSAMI is a web platform developed by Fraunhofer AICOS to perform analytics on user interaction, by collecting data on this interaction and analysing it with analytical algorithms.

7

Actuator Sub-System: The Auditory Cueing

**Vânia Guimarães[1], Rui Castro[1], Angels Bayés[2], Carlos Pérez[3]
and Daniel Rodriguez-Martin[3]**

[1]Fraunhofer Portugal AICOS – Assistive Information and Communication
Solutions, FhP-AICOS, Porto, Portugal
[2]Centro Médico Teknon – Grupo Hospitalario Quirón, Parkinson Unit,
Barcelona, Spain
[3]Universitat Politècnica de Catalunya – UPC, CETpD – Technical Research
Centre for Dependency Care and Autonomous Living,
Vilanova i la Geltrú (Barcelona), Spain

7.1 Introduction

Patients with Parkinson's Disease (PD) are usually affected by clinical symptoms associated to a reduced motor performance that frequently compromises
their ability to walk independently and safely [1]. While a number of pharmacological solutions help to manage the symptoms of PD, some gait-related
problems appear resistant to such treatments and, over time, movement-related
disturbances turn out to be the most incapacitating symptoms of the disease [2].

The internal regulation of step length is generally affected, which is
reflected as an inability to generate sufficient amplitude of movement, even
though the control of cadence, or step rate, is intact. This is in fact the
foundation for the use of external rhythmic stimuli, or cueing, as a way to
ameliorate the impact of the disease on rhythmic movement-based tasks, as is
the case of walking [3, 4].

Typically, to compensate for the reduced step length, patients may increase
the stepping frequency. These symptoms are usually referred to as "*continuous
gait problems*", since the changes in walking patterns are more or less
consistent from one step to the next. With disease progression, "*episodic
gait disturbances*" may appear which result in intermittent and occasional
episodes, such as start hesitation, Freezing of Gait (FOG) episodes and
festination [3–5].

Several cueing strategies, including acoustic and visual cues, can be applied being capable of modifying movements' speed and amplitude, as well as reducing or shortening the occurrence of episodic gait disturbances, such as FOG episodes [2, 6, 7].

Rhythmic auditory stimulation, in particular, enable people to involuntarily synchronize the rhythm of the steps with the rate of the sounds, therefore, enhancing the sense of "taking a step", and the sense of "rhythm", which are both affected by the disease [2, 7–10].

Multiple studies have found a positive impact of the rhythmic sounds on steps length [8, 11, 12], walking speed [9, 10, 12, 13], variability [3, 9] and FOG episodes [2, 14–16]. In addition to the positive impact observed on gait, it was found that rhythmic sounds can also enhance the performance of other tasks involving perceptual and motor timing [17].

Many external cueing approaches are described in the literature; however, several limitations can be identified. Current solutions typically provide interventions continuously (i.e., even when no specific gait symptoms are present), and require cueing to be triggered manually, which limits its applicability in people's everyday life [2]. Also, the application of a continuous cueing means that stimuli are presented without considering whether the patient is suffering from walking problems or not, which can lead to a possible habituation effect that reduces cueing efficacy [18].

To address some of the limitations encountered on previous approaches, an automatic Auditory Cueing System (ACS) was developed throughout the project REMPARK. The developed solution used a smartphone and a headset available in the market, and could provide external cues automatically each time a relevant motor symptom was detected requiring cueing activation. For that purpose, the REMPARK sensor sub-system was used.

This chapter is focused on the Auditory Cueing System (ACS) developed for REMPARK as part of the project implementation. Last part of the chapter is devoted to a technical discussion on the possible future use of the REMPARK developed sub-systems (mainly, the sensor) for an automatic control of the dosage of drug (apomorphine) delivered by an infusion pump, when integrated into the reaction loop around the patient. This technical validation opens the door to future experiences in the way to a better management of the PD symptoms, meaning an improvement on patients' quality of life.

7.2 Cueing Strategies for Gait in Parkinson's Disease

Human beings are particularly sensitive to the temporal characteristics of sound, therefore, *sonification* (generation of data-dependent audio to present information) suits well for time-related tasks, as is the case of the body movement. Other attributes besides rhythm can also be associated to certain events or processes by listening. Pitch (i.e., the perceptual dimension of frequency), for example, is related to the perceived urgency of a warning and, the higher the pitch, the higher the perceived urgency [19].

Multiple types of sounds were explored by different authors as strategies to guide and provide feedback on walking patterns in people with PD, including, metronome and music. In all strategies employed, rhythm turns out to be a key aspect in influencing time-related tasks, as is the case of walking.

The concept of metronome is generally applied to a device that produces regular, metrical beats with adjustable number of beats per minute (BPM). It is typically used by musicians as a reference to help keeping a steady tempo and was also applied in the context of cueing in PD.

The metronome generates temporal expectations that can be intuitively associated to the cyclic movement of walking. In fact, it allows people with PD to predict when the next step should occur, which facilitates movement optimization and execution [17]. Therefore, it is required that an adequate rhythm is provided to the user, considering his/her normal walking cadence.

The metronome can pace both – right and left – footfalls, by producing a regular, repeated sound at an adjustable pace defined as the number of beats per minute (BPM). Several authors showed that the presentation of sounds with a rhythm that is lower than their natural walking rhythm can be used not only to influence rhythm but also to indirectly influence the magnitude of the movement through an increase in the length of the steps, while maintaining gait speed [9, 20]. A positive impact was also observed in other walking parameters, including the variability of walking [3, 9] and FOG episodes [2, 14–16]. Rhythmic sounds could also enhance the performance of other tasks involving perceptual and motor timing [17].

More complex sounds, including music, have also shown to influence the organization of movement in time and space. Music, like movement, is multidimensional, as such it can be naturally linked to spatial, temporal and force elements of the movement. It can be used as an immediate entrainment stimuli acting during movement, or a facilitating stimulus for training to achieve more functional gait patterns. Music therapists are trained to adjust rhythm, dynamics, and pitch as needed specifically by each patient, according to their rehabilitation needs [21].

Wittwer, et al. in [22] showed that music produced a significant increase in gait velocity in healthy older adults, due to a significant increase in stride length. In contrast, when applied to a group of people with Huntington's disease, and contrarily to what would have been expected, music produced no results, and, in contrast, metronome performed better in this group.

The authors suggested that participants' cognitive deficits may have impaired their ability to discern the beat from the more complex music structure [22].

Besides auditory feedback, other types of external cueing can be applied, including visual and somatosensory cues. Visual cueing is traditionally employed using series of strips placed on the floor in transverse line for the patients to walk over. However, this type of strategy can only be applied in laboratory context, not being useful during the daily life. Portable solutions, such as goggles with light emitting diode (LED) or laser-guided walking canes have been developed, in which the same principles of projecting parallel lines on the floor are applied (see Figure 7.1). Some of these solutions can even present the stimuli on demand, for example, when a FOG episode occurs. However, some issues arise from the use of this kind of strategies in outdoor environments, especially in bright areas where the visual information may become less visible. Moreover, this kind of assistive devices may not be practical for real life usage conditions, due to its obtrusiveness [18, 23].

In addition to these strategies, also rhythmic somatosensory cueing has been explored. Strategies such as electrical stimulation or rhythmic vibration have been studied by different researchers, showing also positive effects on gait [23]. Electrical stimulation, in particular, requires more complex setups that may not be practical for a daily usage [25].

In [23] it is shown that although all the three types of stimuli (visual, auditory, somatosensory) were effective in improving gait velocity, step length and cadence, auditory cueing was in fact the most effective cueing strategy applied. Moreover, it may provide the easiest setup, being more practical and realist for a daily everyday usage.

Figure 7.1   Examples of visual cueing strategies (extracted from [24]).

7.3 State of the Art: Competitive Analysis

A systematic search was conducted on the World Wide Web in order to capture and analyse relevant information about existing studies, projects and commercial solutions, with a particular focus on solutions for auditory stimulation and feedback in PD.

Research included technology being developed and described in published and publicly available papers and conference proceedings, as well as commercial solutions that are already available in the market. Moreover, a patent search was conducted to find out innovative solutions for the management of PD. Search was limited to solutions targeting patients with movement disorders, in particular, patients with PD and technological approaches mostly based on auditory cueing aiming to reduce or overcome motor symptoms related to the disease.

Table 7.1 summarizes the results of said search for R&D and market-ready products. For each solution, the following parameters were evaluated:

- Current stage of development: R&D or market-ready;
- System Components: e.g. smartphone, headset, dedicated hardware, etc.;
- Types of cueing: metronome, music, verbal cueing and possibly others (e.g. visual or sensory cueing);
- Modalities and symptoms: identifies whether a system can or not provide cueing automatically in response to motor symptoms (i.e. the actuation mode); also identifies whether cueing can or not be manually activated under specific circumstances/requirements with the purpose of training (i.e. the training mode); provides a list of motor symptoms that are targeted by the system;
- History Recording, in case the system is able to record the history of activations/deactivations;
- Connection with a Disease Management System (DMS), in case the system is prepared to connect to a server, in which data are recorded to be then displayed and used by clinicians.

Not many non-pharmacological solutions for actuation in PD can currently be found in the market. The majority of the solutions found are still in R&D stage. In particular, related to the auditory stimulation, only one commercial solution was found, the GAITAID Virtual Walker. This solution can provide not only auditory stimulation, but also visual stimulation and for that it requires the use of display glasses with built-in earphones, which are connected to a proprietary control unit that is responsible for creating images and sounds by responding to the user's movements. The system was made in the US, and the

Table 7.1 Technology watch on actuators for auditory cueing in Parkinson's Disease

Solutions	Development Stage	Components	Type of Cueing				Features – Modalities and Symptoms			Other Features	
			Metronome	Music	Verbal Cueing	Others	Actuation Mode	Training Mode	Motor Symptoms	History Recording	Connection w/DMS
REMPARK ACS [30]	R&D	Android Smartphone, Headphones, Sensor	✓	×	✓	×	✓	✓	Bradykinesia, Dyskinesia, FoG	✓	✓
GAITAID Virtual Walker [32]	Market-Ready	Control Unit, Display glasses with built-in earphones	✓	×	×	Visual cueing	×	✓	×	×	×
iRACE [33]	R&D	iOS-based mobile device	✓	×	×	×	×	✓	×	✓	✓
GaitAssist [34] (Project: CuPID [35])	R&D	Two body-worn IMU sensors, Smartphone, earphones	✓	×	×	×	✓	✓	FoG	×	×
CuePack [36] (Project: RESCUE)	R&D	Control Unit, Display glasses and earphones	✓	×	×	Visual and sensory cueing	×	✓	×	×	×
Bächlin, et al. [2]	R&D	Multiple sensors, Processing Unit and earphones	✓	×	×	×	✓	×	FoG	×	×

prices range from 1995 USD (~1840 €), for orders in US, to 2145 USD (~1980 €), for orders outside US and Canada. GAITAID, unlike the REMPARK solution, is not capable of detecting specific motor symptoms, neither actuating automatically upon the detection of the symptoms, being therefore more limited in this sense.

In the study develop by Bächlin, a system for the automatic detection of FOG episodes was developed. For that, they require the use of multiple sensors, which are placed on the waist, thigh and ankle. The system is controlled by a portable processing unit which is placed on the waist and is responsible for processing sensor data and triggering auditory cueing (metronome sound) as soon as a FOG episode is detected. It clearly demonstrates the positive and immediate effects of auditory stimulation in patients' gait.

The iRACE application, developed for iOS, uses the sensors of the smartphone to detect steps and estimate steps length. The finger tapping test is also available in the application. The GaitAssist, which was developed in the FP7 European project CuPID, also uses the smartphone as a support for the execution of physical exercises for training, and enable the detection of FOG episodes through the analysis of the data coming out of two sensors which are placed in the ankles.

CuePack, which was developed in the context of the European project RESCUE, enables the application of different types of stimulus, including auditory, visual and sensory stimulation. Its actuation capabilities are limited to a training context, since it is not able to detect motor symptoms.

According to Table 7.1, no cueing device is as complete as REMPARK ACS. While some solutions are not capable of detecting motor symptoms and actuating accordingly, others, are limited to the detection of FOG episodes. Also, to detect this type of episodes, both solutions require the use of multiple sensors, whereas REMPARK needs just one sensor unit to detect a larger amount of motor symptoms, including FOG episodes, bradykinesia and dyskinesia events.

Some of the presented solutions are based on the use of specific equipment for sounds generation and control, as is the case of GAITAID Virtual Walker, Bächlin and CuePack. In contrast, REMPARK, GaitAssist and iRACE are based on the use of a smartphone.

Moreover, REMPARK can connect to a DMS, enabling also clinicians to analyze their patients' data. As expected, and since the majority of the systems are still in the R&D stage, no references to the connection to a server are usually available in the considered systems.

It was noticed that, to actuate properly, the majority of the systems aforementioned, including REMPARK, require an initial configuration step, in which normal gait parameters for a specific patient with PD are identified and given as an input to the system. This step helps not only to decide when a gait pattern deviates from the normal, but also to establish the functional goals that need to be achieved after cueing is applied.

Regarding the type of sounds delivered to the patients, the majority of the solutions are based on the metronome, i.e. repetitive sounds played with a certain periodicity. REMPARK also offers the possibility of delivering verbal cueing (i.e. "one-two-one-two"). Additional solutions are also capable of providing other types of cueing. CuePack, for example, is able to provide three different types of cueing, including the metronome, visual cueing and sensory cueing. GAITAID Virtual Walker can provide both auditory and visual cueing, which can be used in combination or separately, according to the patient specificities. However, for this functionality, it requires the use of proprietary devices that need to be carried by the patient while walking.

Based on this analysis, and considering the limitations of the current solutions encountered on the literature, REMPARK seems to be the most integrated solution for detection, treatment and management of motor disorders in PD, being capable of actuating in real time, each time a motor symptom is detected and cueing is required. Moreover, it can be used either as a gait training device (at clinics or at home) or during the daily life, at home or outdoors, being capable of actuating automatically when a motor symptom is detected and cueing is deemed as required. As it uses mainly components that already exist in the market, such as the smartphone and a headset, and a single discrete sensor that needs to be worn in the belt, the solution itself may not be stigmatizing for its users, being therefore more attractive.

A more throughout analysis of the existing technology was conducted through the analysis of the patents produced in this area. The most relevant inventions are described in the following paragraphs:

- US 8409116 B2 (Granted, filing date 2010) presented a device and method for treating patients with movement disorders. The device includes a sensor for detecting an akinetic episode (e.g. a FOG episode) of the person and a receiver that automatically issues a single or multiples cues to aid in the restoration of the movement. The cue can be of any type, including verbal, auditory, physical or tactile. Moreover, they propose sending the information about FOG events wirelessly to a caregiver and/or an emergency contact [26].

- EP 2346580 A2 (Application, filing date 2011) disclaims a method for the improvement of gait parameters comprising determining attributes of ideal spatial and temporal gait parameters of the subject, measuring the actual parameters and determining the rhythmic audio cue that may cause the subject to improve a gait parameter [27].
- US 8961186 B2 (Granted, filing date 2012) presents an ambulation accessory that can also be used as a training device. The invention includes two balls that illuminate in an alternating fashion to provide a visual target to each leg. When the user reaches the ball with his foot, the system returns visual and auditory feedback indicating a successful step [28].
- WO 2012177976 A2 (Application, filing date 2012) presents a method performed by one or more processing devices that includes generating a visual representation of an object in the environment, retrieving an auditory stimulus indicative of the location of a virtual target in the environment, determining the proximity of the object to the virtual target and adjusting the auditory attributes based on proximity [29].

As can be perceived, some of these patents describe a system whose operating mode is basically the same adopted by the systems in Table 7.1. They include the possibility of detecting movement disorders, the ideal gait parameters and the cueing strategy/parameters required to overcome said disorder. All the systems described in the inventions offer the possibility of using auditory cueing as a feedback mechanism for gait. US 8409116 B2 presents a generic method that can use just one type of cueing or a combination of several types. EP 2346580 A2, on the other side, is focused on the use of audio cues comprising a tone and a beat. The last two commented patents (US 8961186 B2 and WO 2012177976 A2) are mainly focused on the use of visual cues combined with audio for feedback on gait. As can be understood, these systems lack practicability, as the delivery of visual stimulus while walking requires the use of additional equipment, e.g. display glasses. Systems based on the use of auditory cueing can be largely simpler and affordable, since they can benefit from the use of existing, commercial, equipment that a person may already have, as is the case of REMPARK ACS using standard headsets and a smartphone.

7.4 REMPARK Auditory Cueing System (ACS)

The ACS was developed in the context of the project REMPARK according the specifications and characteristics described in Chapter 3. It uses an Android smartphone, which is connected via Bluetooth to a commercial headset.

The smartphone is responsible for generating cueing sounds, controlling their rhythm and streaming them to the headset (Figure 7.2).

The headset can be any commercial Bluetooth headset with, at least, Bluetooth version 2.1 and A2DP profile [30].

The ACS includes multiple types of sounds, which can be chosen through the smartphone interface. It includes metronome sounds, musical beats, clapping and verbal cueing ("One-two-one-two"). The patient may himself select the sound that he/she prefers the most and is more comfortable with.

Each sound, in reality, falls into the category of a metronome, considering that all sounds produce metrical beats with adjustable number of beats per minute (BPM). The rhythm of the sounds is provided in beats per minute for a direct connection with gait cadence or step rate, also defined as the number of steps per minute (SPM). Both footfalls, right and left, are paced by the system (Figure 7.3), and some of the available sounds are actually a combination of different timbres that aim at providing a different feedback for each step.

The ACS takes advantage of the existing modules that are integrated in the REMPARK solution, including the motor symptoms detection provided by the REMPARK sensor sub-system. The auditory cueing is activated by the system as soon as FOG or bradykinesia are detected by the movement sensor. Once a patient has continued walking without bradykinesia or has stopped walking, the cueing is discontinued. Nevertheless, each time auditory cues are activated, a pop-up window appears on the smartphone screen, for a quick

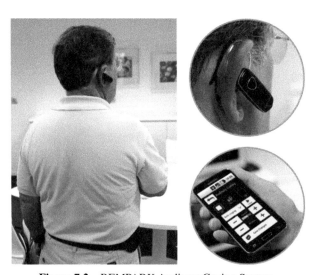

Figure 7.2 REMPARK Auditory Cueing System.

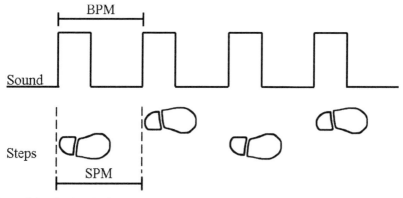

Figure 7.3 Cueing rhythm (beats per minute BPM) and cadence (steps per minute SPM).

interaction with the system, which enables the patient to easily stop the sounds, or manipulating sounds rhythm and volume (Figure 7.4).

The rhythm and volume of cues are pre-configured through the smart-phone. A clinician may introduce the target rhythm for the cueing sounds that will target patients' needs the best as possible. The clinician may also define the maximum and minimum cueing rhythm that can be provided to

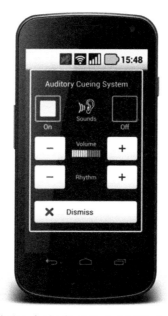

Figure 7.4 Pop-Up window for basic control of ACS when it starts automatically.

the patient, so that rhythm may never compromise patients' safety and the safe synchronization of steps with gait cueing, while still maintaining a good pattern of walk. The clinician may also help the patient define the minimum volume accepted by the system, so that it may never go below a level that is inaudible by the patient.

Voice instructions are also available in the system, to explain what to do, or how to proceed each time sounds start playing. Both temporal instructions ("Step in time to the beat") and spatiotemporal instruction ("Take a big step in time to the beat") can be delivered according to the type of symptoms detected. These instructions can be deactivated through the ACS settings application. Through this screen, the patient may also preview each sound available, as well as change the default actuator type of sound.

The ACS can also be activated manually by the patient, and therefore serve as a training device for gait. This will enable patients to listen to the sounds as part of a gait training task. The controller application can be opened from the REMPARK applications list (Figure 7.5).

Moreover, the ACS takes advantage of the capability of recording the history of activations and deactivations, which can then be analyzed by clinicians through the disease management system (DMS) [30].

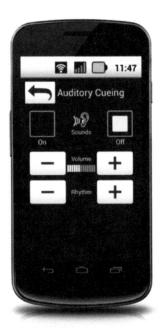

Figure 7.5   Auditory Cueing System application.

7.5 Outcomes from Field Trials: Future Considerations

The ACS was developed as a module capable of providing auditory stimuli in real time, to help people with Parkinson's Disease improve speed and amplitude of walking. Before developing the final version of the system, which was then evaluated in the 3 days-long trials stage, a preliminary testing stage with a first prototype was required to get the first impressions on the feasibility of the system in stimulating gait, as well as evaluate system's usefulness and acceptance.

Preliminary tests were conducted in Spain with 12 people with PD. During the tests, patients were asked to walk, sometimes with cueing, other times without cueing, and their walking rhythm was measured against cueing rhythm. Motor symptoms were identified by observation by a trained medical doctor or therapist. Patients were asked to walk along predefined distances and walking circuits, but also to walk at will, being able to choose the trajectory they wanted to follow. Therefore, both indoor and outdoor conditions were captured (Figure 7.6).

Despite the reduced number of participants in this preliminary testing stage, experimental results suggested that better gait patterns could be stimulated when individuals follow rhythmic sounds whose rate is similar to their natural step rate. The fact that people tended to walk with better walking patterns and overcome gait problems when feedback rhythm was closer to their normal walking cadence supports this observation.

Figure 7.6 Preliminary tests with PD patients for the ACS.

Additionally, people found the volume and quality of sounds adequate, even in outdoor conditions, where a louder environmental noise could be found. This is an important requirement for the system, since it is expected to work in real time, even in outdoor conditions during a person's daily routine. Actually, all participants would be willing to use the system during their everyday life, considering that it would help them in real time during their daily activities. A complete report of methods and results achieved in these preliminary tests can be found in reference [31].

After the final ACS prototype has been developed and integrated with the other REMPARK components, i.e. the REMPARK sensor and the DMS, the final trials period took place, as it is reported in Chapter 9. During the trials, 41 people with PD used the system continuously for three consecutive days and the ACS was put to the test under real life conditions.

After using the system continuously during three consecutive days, people reported some issues related to the size and comfort of the headset. In fact, the headset chosen for the trials was an existing commercial device that was chosen considering mainly its technical specification and price. However, for a final solution, it is also required to consider its aspect, weight, easiness to put and remove from the ear, ear fitting, sense of attachment and sound quality, and take into account also the individual preferences of its users.

In addition, some patients suggested the use of music instead of the metronome to act as cueing during their daily life, to be more enjoyable and not so monotonous. In fact, due to the requisite of maintaining a fixed rhythm, the metronome can become monotonous after a while, therefore challenging the acceptability and sustainability of the solution. Still, it can be the only viable cueing solution for some people, for example, people with cognitive deficits, as suggested by the authors in reference [22].

As a final conclusion, the REMPARK system appears to be usable and participants seem to be satisfied with the system. Improvements on the system will, for sure, take into account the valuable input provided by the people who participated in the trials.

7.6 A Step Further in REMPARK: Automatic Drug Administration

As it has been already discussed in Chapter 3, one of the project goals was to technically validate the feasibility of using REMPARK system to create an automatic or semi-automatic feedback wireless control loop that would improve the treatment of PD patients, using some reaction solutions. For this purpose, a commercial apomorphine infusion pump was modified. The modifications consisted of incorporating the necessary electronic components so that it is able to communicate with the smartphone in order to receive commands related to the drug dosage to be administered. These commands come from the symptoms' measurements performed by the developed REMPARK sensor. Additionally, the electronic system added to the pump should permit a manual user interaction through the smartphone.

REMPARK project proposal discarded, from the beginning, the organization of a specific pilot action using drug delivery systems control with real patients because this was out of the real possibilities of the project piloting activity. The main reason was the required timing for these type of pilots, where time scales are much more long than those for the rest of scheduled assessment pilots. The only activity, at this level, was to consider the effective possibility of a wireless control of these devices, preparing all the requirements and conditions from a technical and functional point of view.

Apomorphine has been widely used to treat Parkinson's Disease. One of its main usages is under Apo-go Pen format. Doses are small injections which patients can self-administer, with a limited amount of apomorphine. Its active principle provokes, compared to the levodopa effect, a very quick response (within few minutes), and the length of the effect is short, which makes this drug suitable for controlling the symptoms in real time.

At that level, REMPARK project proposed several objectives:

- to technically demonstrate that an automatic drug delivery system can be remotely controlled and activated by the REMPARK system when an unexpected OFF period is detected.
- an automatic delivery of the rescue dose is technically possible.

A subcutaneous infusion pump would be included in the system and would be ready for its automatic activation. This prototype will facilitate functional laboratory testing of the drug delivery pump for remote operation in simulated clinical conditions.

After analysing the existing options, the Microjet CronoPAR [37] subcutaneous pump manufactured by Canè s.r.l was identified as the unique option to be adopted. This commercially available pump provides the functional features associated with the infusion of apomorphine to Parkinson's Disease patients. In order to fulfil the project requirements, an electronic control unit was developed and integrated into the device. Figure 7.7 shows this infusion pump model as it is commercially available.

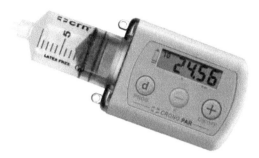

Figure 7.7   Infusion pump to be adapted for a possible use with REMPARK system.

In the original pump, a set of buttons enable the user to command the behaviour of the pump. A display shows the state of the device. As the device does not have a battery recharge system, the battery must be replaced and it may be accessed removing a side tap.

An automatic control requires the substitution of the manual interaction by an electronic interface, with the equivalent functionality and provided with a connection to the smartphone using a wireless communication channel. The strategy for this modification was to replace the PCB (Printed Circuit Board) integrated in the pump keypad by a much more complex PCB integrating the interaction with the pump, through the electronic control of the buttons, and the mentioned wireless connection.

Figure 7.8 shows the modified Microjet CronoPAR, where the controller has been replaced by a new electronic hardware (PCB). The PCB was designed to allow a similar aspect of the device, with the same functionality and its operative LCD display.

The main directive in the PCB design process was the achievement of a device with reasonable autonomy but keeping the original physical size, allowing the user to wear the subcutaneous pump without undue inconveniences. The components of the system were restricted to be compatible with an adjusted power consumption and practical size. Figure 7.9 shows the general scheme of the PCB device. Internally, the device includes a microcontroller that handles the different sub-system parts. The microcontroller provides the system with the capacity of managing the subcutaneous pump and at the same time to establish a wireless communication.

The control board of the PCB is managed by a microcontroller, which is in charge of controlling the subcutaneous pump while managing the data received from the wireless module. The PCB replaces the original control board of the subcutaneous pump maintaining the same shape, but increasing its thickness.

Figure 7.8 Modified pump.

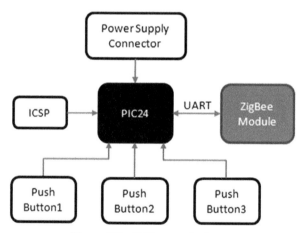

Figure 7.9 PCB general scheme.

In order to control the pump behaviour, the control buttons have been replaced by control signals provided by the PCB. As a feedback, to confirm the proper operation of the pump, an electronic signal associated with internal acoustic information is provided.

The Microjet CronoPAR is supplied with a standard and commercially available 3 Volt type 123 A Lithium non-rechargeable battery with a capacity of 1400–1500 mAh. As the operating voltage range of the PCB is 2.1 V to 3.6 V, it can be supplied with the battery of the original pump. This feature allows

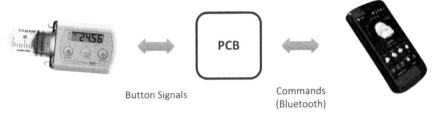

Button Signals   Commands
                                  (Bluetooth)

Figure 7.10   System operation overview.

the PCB to work with all the available operating voltage ranges, ensuring that the system can work with any low battery situation.

The control and supervision functionalities of the modified pump are achieved with the execution of the microcontroller's embedded firmware. The system operation relies on a series of independent but coordinated processes that manage the behaviour of the subcutaneous pump, taking advantage of its functional capabilities. The PCB device takes control of the pump behaviour working as a wireless interface between the subcutaneous pump and the smartphone (Figure 7.10). The main state machine (firmware) is in charge of coordinating the commands received from the communication module, giving the necessary functionality to the system and synchronizing the remaining processes. Moreover, the subcutaneous pump provides multiple features (e.g., administration of the bolus dose). These features have been developed as a secondary software (state machine), which is managed by the main firmware software of the device.

It is necessary to confirm that the control of the pump is correctly performed, and also, it is necessary to acknowledge and monitor the correct operation of the device. In consequence, an interface was provided giving the necessary information to the system. This information about the functioning of the pump is provided by means of an electronic signal, associated with an internal acoustic confirmation. Using this signal to complete the feedback loop, the entire system can be managed electronically. As a result, the PCB interface is connected to the subcutaneous pump through four signals: three output signals connected to the buttons of the pump, and one input connected to the beep signal. The connecting diagram between the subcutaneous pump and the PCB can be seen in Figure 7.11.

In order to manage and control the pump, the input signals generated by the front buttons are now provided by the PCB. Then, by activating these signals, the PCB is able to perform all the required actions, for example administer bolus dose or program the infusion rate.

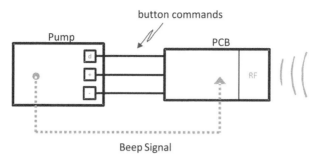

Figure 7.11   Connection diagram between the pump and the designed PCB.

In normal operation regime, the ambulatory infusion pump manufactured by Canè s.r.l. offers two main operative modes: intermittent bolus or continuous apomorphine infusion at variable dose. When operating in a single shot mode the subcutaneous pump administrates a given quantity of drug only once. On the other hand, continuous apomorphine infusion consists on continuous pulses of infusion, where the rate of these pulses depends on the programmed infusion rate.

In REMPARK, the implemented firmware on the PCB is trying to replicate this operative in an electronic way, using a state-machine approach. Deployment of pump behaviour in state-machine based allows a good understanding of the logic process. In principle, the subcutaneous pump has two independent states; *Pump Switched ON* and *Pump Switched OFF*. Each state has the possibility to perform specific actions. For example, the pump must be switched off to program the bolus dose. On the other hand, switching on the pump allows to administer infusion, bolus dose and to program the flow rate.

The OFF state starts when the pump finishes an auto diagnostic routine. Here the bolus dose configuration may be managed. On the other hand, infusion state starts when the pump has been switched on. This state is in charge of managing the main actions that can be requested remotely and mainly the administration of the infusion and the bolus dose. The PCB interprets the command sent remotely and changes the state of the pump in consequence.

To ensure that actions are correctly performed, a feedback from the pump is obtained, as it was introduced. The pump confirms all its actions by emitting a brief sound signal. Then, taking this signal as an input, the PCB is able to read and analyse the action feedback. As a main characteristic, this sound signal has different behaviours depending on the action performed by the pump. For example, the length of the beep signal when the pump changes to infusion mode is different from that generated when it changes to bolus dose administration mode. Reading and processing this signal by the PCB allows the identification and confirmation of the action performed by the pump at any time.

As it has been already explained, the aim of REMPARK was not to include experience with patients around this implementation of an automatic control loop including the apomorphine pump, but just to demonstrate the technical viability and its feasibility.

The inclusion of a wirelessly controlled infusion pump in the REMPARK system would facilitate some future scenarios:

- If on-line symptoms recognition is done correctly, it opens the door to a better administration of the medication doses, since dosages would be automatically adapted to the immediate needs of patients.
- Usually the practitioner has difficulties for adjusting a suitable dose for the pump's continuous mode, and also to control extra doses administered by the pump. REMPARK can help the professionals to properly assess the number of OFF hours that a given patient has experienced with the aim of judging the effect of the pump therapy through an objective information.
- Usability from the patient's point of view. Some patients with Parkinson's have OFF phases so severe that they cannot even self-administrate extra doses. Patients with severe OFFs, which have no caregivers who can perform this task for them, often cannot choose the treatment with continuous infusion pumps. REMPARK could contribute to improve the quality of life of these patients.

7.7 Conclusion

Current chapter has presented the REMPARK activity on PD actuation as a possible improvement for a more effective management of the disease symptoms.

Auditory Cueing strategy is already a well-known solution for the improvement and facilitation of a better gait of people with Parkinson's when invalidating symptoms appear. REMPARK tries to improve this technique with some new characteristics and possibilities derived from an effective detection of symptoms on-line during the normal activities of the patients.

In a more speculative scenario, REMPARK has demonstrated the technical viability and the feasibility of an electronic control to be added to an infusion pump of apomorphine, allowing in the future, an automatic delivery of the drug depending on the patient specific needs.

References

[1] M. E. Morris, F. Huxham, J. McGinley, K. Dodd e R. Iansek, "The biomechanics and motor control of gait in Parkinson disease," *Clinical biomechanics,* vol. 16, no. 6, pp. 459–470, July 2001.

[2] M. Bächlin, M. Plotnik, D. Roggen, N. Giladi, J. M. Hausdorff e G. Tröster, "A wearable system to assist walking of Parkinson's disease patients," *Methods of information in medicine,* vol. 49, no. 1, pp. 88–95, January 2010.

[3] J. M. Hausdorff, "Gait dynamics in Parkinson's disease: common and distinct behavior among stride length, gait variability, and fractal-like scaling," *Chaos (Woodbury, N.Y.),* vol. 19, no. 2, pp. 1–14, June 2009.

[4] M. F. del Olmo e J. Cudeiro, "Temporal variability of gait in Parkinson disease: effects of a rehabilitation programme based on rhythmic sound cues," *Parkinsonism & related disorders,* vol. 11, no. 1, pp. 25–33, January 2005.

[5] N. Giladi, H. Shabtai, E. Rozenberg e E. Shabtai, "Gait festination in Parkinson's disease," *Parkinsonism & Related Disorders,* vol. 7, pp. 135–138, 2001.

[6] M. Rodger, W. Young e C. Craig, "Synthesis of Walking Sounds for Alleviating Gait Disturbances in Parkinson's Disease," *IEEE Transactions on Neural Systems and Rehabilitation Engineering,* vol. 22, no. 3, pp. 543–548, May 2014.

[7] M. P. Ford, L. A. Malone, I. Nyikos, R. Yelisetty e C. S. Bickel, "Gait Training With Progressive External Auditory Cueing in Persons With Parkinson's Disease," *Archieves of Physical Medicine and Rehabilitation,* vol. 91, no. 8, pp. 1254–1261, August 2010.

[8] A. Picelli, M. Camin, M. Tinazzi, A. Vangelista, A. Cosentino, A. Fiaschi e N. Smania, "Three-dimensional motion analysis of the effects of auditory cueing on gait pattern in patients with Parkinson's disease: a preliminary investigation," *Neurological sciences,* vol. 31, no. 4, pp. 423–430, August 2010.

[9] K. Baker, L. Rochester e A. Nieuwboer, "The effect of cues on gait variability – reducing the attentional cost of walking in people with Parkinson's disease," *Parkinsonism & related disorders,* vol. 14, no. 4, pp. 314–320, January 2008.

[10] M. Suteerawattananon, G. S. Morris, B. R. Etnyre, J. Jankovic e E. J. Protas, "Effects of visual and auditory cues on gait in individuals with Parkinson's disease," *Journal of the Neurological Sciences,* vol. 219, no. 1–2, pp. 63–69, April 2004.

[11] W. Nanhoe-Mahabier, A. Delval, A. H. Snijders, V. Weerdesteyn, S. Overeem e B. R. Bloem, "The possible price of auditory cueing: Influence on obstacle avoidance in Parkinson's disease.," *Movement disorders,* vol. 0, no. 0, pp. 1–5, February 2012.

[12] S. Ledger, R. Galvin, D. Lynch e E. K. Stokes, "A randomised controlled trial evaluating the effect of an individual auditory cueing device on freezing and gait speed in people with Parkinson's disease," *BMC neurology,* vol. 8, no. 46, January 2008.

[13] T. Howe, B. Lovgreen, F. Cody, V. Ashton e J. Oldham, "Auditory cues can modify the gait of persons with early-stage Parkinson's disease: a method for enhancing parkinsonian walking performance?," *Clinical Rehabilitation,* vol. 17, no. 4, pp. 363–367, July 2003.

[14] J. Spildooren, S. Vercruysse, P. Meyns, J. Vandenbossche, E. Heremans, K. Desloovere, W. Vandenberche e A. Nieuwboer, "Turning and Unilateral Cueing in Parkinson's Disease Patients with and without Freezing of Gait," *Neuroscience,* vol. 207, pp. 298–306, 2012.

[15] Z. Kadivar, D. Corcos, J. Foto e J. Hondzinski, "Effect of step training and rhythmic auditory stimulation on functional performance in Parkinson patients.," *Neurorehabilitation and Neural repair.,* vol. 25, no. 2, pp. 626–635, 2011.

[16] R. Velik, "Effect of On-Demand Cueing on Freezing of Gait in Parkinson's Patients," *International Journal of Medical, Health, Pharmaceutical and Biomedical Engineering,* vol. 6, no. 6, pp. 8–13, 2012.

[17] C.-E. Benoit, S. D. Bella, N. Farrugia, S. Mainka e S. Kotz, "Musically cued gait-training improves both perceptual and motor timing in Parkinson's Disease," *Frontiers in Human Neuroscience,* vol. 8, July 2014.

[18] R. Velik, U. Hoffmann, H. Zabaleta, J. F. M. Massó e T. Keller, "The effect of visual cues on the number and duration of freezing episodes in Parkinson's patients," *Proceeding of the 34th Annual International Conference on Engineering in Medicine and Biology Society,* 2012.

[19] J. Forsberg, "A mobile application for improving running performance using interactive sonification," Stockholm, Sweden, 2014.

[20] E. van Wegen, C. Goede, I. Lim, M. Rietberg, A. Nieuwboer, A. Willems, D. Jones, L. Rochester, V. Hetherington, H. Berendse, J. Zijlmans,

E. Wolters e G. Kwakkel, "The effect of rhythmic somatosensory cueing on gait in patients with Parkinson's disease," *Journal of the Neurological Sciences,* vol. 248, no. 1–2, pp. 210–214, October 2006.

[21] R. F. Pfeiffer, Z. K. Wszolek e M. Ebadi, Parkinson's Disease, 2nd ed., T. &. F. Group, Ed., CRC Press, 2013.

[22] J. E. Wittwer, K. E. Webster e K. Hill, "Music and metronome cues produce different effects on gait spatiotemporal measures but not gait variability in healthy older adults," *Gait and Posture,* vol. 37, no. 2, pp. 219–222, February 2013.

[23] C. Pongmala, A. Suputtitada e M. Sriyuthsak, "The study of cueing devices by using visual, auditory and somatosensory stimuli for improving gait in Parkinson patients," *2010 International Conference on Bioinformatics and Biomedical Technology (ICBBT),* pp. 185–189, 2010.

[24] M. Defranco, "Parkinson's Tips for Freezing of Gait from our Physical Therapist," Center for Movement Disorders and Neurorestoration, 23 November 2011. [Online]. Available: http://movementdisorders.ufhea lth.org/2011/11/23/physical-therapy-tips-for-freezing-of-gait/. [Acedido em September 2016].

[25] G. Lyons, T. Sinkjaer, J. Burridge e D. Wilcox, "A Review of Portable FES-Based Neural Orthoses for the Correction of Drop Foot," *IEEE Transactions on Neural Systems and Rehabilitation Engineering,* vol. 10, no. 4, pp. 260–279, 2002.

[26] E. Wang, L. Metman e E. Jovanov, "Method and device to manage freezing of gait in patients suffering from a movement disorder". Patente US 8409116 B2, 2 Abr 2013.

[27] J. Whitall, S. Mccombe-Walker e M. Anjanappa, "Step trainer for enhanced performance using rhythmic cues". Patente EP 2346580 A2, 27 Jul 2011.

[28] P. LoSasso, "Accessory for a walker to improve gait performance". Patente US 8961186 B2, 24 Fev 2015.

[29] D. Polley e K. Hancock, "Auditory stimulus for auditory rehabilitation". Patente WO 2012177976 A2, 27 Dez 2012.

[30] A. Samà, C. Pérez-López, D. Rodríguez-Martín, J. M. Moreno-Aróstegui, J. Rovira, C. Ahlrichs, R. Castro, J. Cevada, R. Graça, V. Guimarães, B. Pina, T. Counihan, H. Lewy, R. Annicchiarico, À. Bayés, A. Rodríguez-Molinero e J. Cabestany, "A double closed loop to enhance the quality of life of Parkinson's Disease patients: REMPARK system," *Innovation in Medicine and Healthcare,* pp. 115–123, 2014.

[31] V. Guimarães, R. Castro, A. Barros, J. Cevada, À. Bayés, S. García e B. Mestre, "Development of an Auditory Cueing System to Assist Gait in Patients with Parkinson's Disease," *Bioinformatics and Biomedical Engineering, Lecture Notes in Computer Science,* vol. 9044, pp. 93–104, 2014.

[32] http://medigait.com/ [Accessed on April 15, 2015].

[33] S. Zhu, R. Ellis, G. Schlaug, Y. Sien Ng e Y. Wang, "Validating an iOS-based Rhythmic Auditory Cueing Evaluation (iRACE) for Parkinson's Disease", MM'14.

[34] S. Mazilu, U. Blanke, M. Hardegger, G. Tröster, E. Gazit e J. M. Hausdorffe, "GaitAssist: A Daily-Life Support and Training System for Parkinson's Disease Patients with Freezing of Gait", Proceedings of the SIGCHI Conference on Human Factors in Computing Systems, pp. 2531–2540, 2014.

[35] http://cupid-project.eu/project [Accessed on April 15, 2015].

[36] http://rescueprojects.org/ [Accessed on April 15, 2015].

[37] CANÈ S.p.A. Medical Technology. Microjet CRONO PAR Manuel d'Instruction. MAN 01/F/04 PARII 09/09. Retrieved from http://www. canespa.it

8

Disease Management System (DMS)

Hadas Lewy

Maccabi Healthcare Services, Tel-Aviv, Israel

8.1 Introduction

Parkinson's Disease is a chronic and progressive neurodegenerative disorder with a great number of motor and non-motor symptoms [2, 4, 5, 7]. The cardinal symptoms are bradykinesia, rigidity, tremor and postural instability [1, 3, 8]. However, a number of non-motor-related symptoms (e.g., sleep disturbances, depression, psychosis, autonomic and gastrointestinal dysfunction as well as dementia) may occur [3–7].

The disease is a great burden as it has a negative incidence on the quality of life, due to a gradual loss of functionality and decreasing ability to take care of oneself. For these reasons, the caring process changes along disease progressions and involves many healthcare professionals. Usually, these professionals are located in different care services and organizations (hospitals and the community), and they cover a wide range of medical care including physicians, nurses, physiotherapists, nutritionists, occupational therapists and social workers. In addition, there is a need to support the family and reduce the informal caregiver burden by providing services, information and knowledge about the treatment.

To cope with the challenging care model of PD patients, the REMPARK project contributes to the development of a Personalized Health System (PHS) with detection, response and treatment capabilities for the remote management of the disease permitting, also, the integration of the generated data with the patient's EHR (Electronic Health Record) and the integration and support of a Disease Management System (DMS), that enable the medical team to provide integrated care and support to the patient and his family.

One of the main problems in most existing systems for disease management is that the data used for treatment is collected on the system overtime

159

and is not updated by other systems that are used for treating the patient. For example, if the patient is treated by the neurologist for the PD and in parallel he is treated by the general physician in the community for other health conditions. Usually, no Integration between data is made and there is a lack of information about the overall condition of the patient. Furthermore, there is no communication between the healthcare professionals treating the patient.

This way of delivering care results in a fragmentation, slight duplication and lack of coordination. The DMS creates an integrated environment for data and knowledge sharing for all care providers without a need to access different organizational systems by users from different organizations and thus overcomes the privacy and security barriers. The DMS also includes clinical guidelines and a decision support tool for healthcare providers that allow the clinical team to obtain accurate and reliable information and to decide about the treatment that best suits the patient for improvement of the disease management.

This tool enables treating several conditions at the same time in a coordinated way; therefore, the organization can use several experts to treat a patient (e.g. psychiatrists, urologists or neurologists) and enable integrated care between different organizations (e.g. hospital community care and welfare) which provides flexibility and efficiency treating a patient with changing conditions, without breaching security constraints of the organization.

For the patient, there is a portal in which he is able to communicate with the professionals (doctor, nurse...) and obtain information about his condition and treatment plan. The DMS communicates with the patient's interface that can be reached by a home computer, a tablet or a Smartphone, it updates the data and collects inputs from the patients, such as self-reported questionnaires, for the evaluation and decision making by the healthcare professionals. These tools give useful information to the patients and empower them for better self-management.

The system also has an interface for healthcare organizations to enable management of the treating teams such as shifts and workload at the call centre, and reports for the management teams.

This chapter describes the use of the implementations of a DMS within the REMPARK system, its ability to communicate with REMPARK sensors, intelligent layer and actuators and reviews the advantages that the DMS brings to all users involved in the system from a clinical, personal (patient), organizational and healthcare points of view. At the end of the chapter, the conclusions from using the system and the vision for its contribution of the future treatment of PD patient are described, as well as for elderly people with co-morbidities in a changing era of care models moving from reactive to proactive treatment, empowering the patient and creating new services and relationship between the patient and the healthcare providers.

8.2 Disease Management System Application

Providing integrated care and remotely treating large populations creates a challenge for the care providers at the clinical and organizational level. Although the data provided by the sensors and analytical tools are valuable and the REMPARK system paves the way for its clinical use in the future for the automatic control of continuous infusion pumps, automatic management of external cues to guide the gait or for comprehensive and reliable analysis of changes in motor state, still the use of this type of data in clinical practice is not implemented. Clinical guidelines for remote monitoring, standardization of care and risk management of the remote monitoring and care is still not well-established and requires development of supportive tools as well as methodological approaches.

The progression of the disease is individual and cannot be predicted, and the patient and their family need support at different levels and by multiple caregivers at different stages of the disease. Since the disease is usually related to elderly people, the associated symptoms may be mistakenly treated as geriatric symptoms and there is a need to diagnose the source of the symptom and provide the personalized treatment considering the overall condition of the patient. The services provided for comprehensive treatment are often provided by different organizations located in the hospital, community service and home care. Therefore, the treatment can be complex and involves several stakeholders.

For these reasons, healthcare system and the industry are interested in the implementation of technologies for remote care in healthcare. For healthcare system, the main challenge is to enable the use of these technologies by changing the model of care and sharing information. Implementation of these technologies requires collaboration of the healthcare professionals and patients not just in adoption but also in the process of development and implementation in new best practice and care pathways and open the way for sharing information from the patient (self-reported) and other sources (different organizations), capturing and analysing patient data from dispersed systems.

For the industry, the challenge is to provide solutions that will support healthcare systems considering constrains of standardization, privacy, reliability, security, existing models of care including healthcare professional workloads and workflows and adapt the solutions to the system [9]. With these challenges in mind, it was proposed and developed the REMPARK system composed of wearable sensors, analytical layer, decision support tool and a care platform (DMS) for professional care providers, patients and family.

The DMS system provides various functionalities that **support the care** by healthcare professionals and communicate with the patient. It is also adaptable to the **organizational care pathways** and **workflows** and flexible to different clinical guidelines that are used in different countries and/or can be developed overtime. It enables to treat the PD patient in a way that changes along the development of the disease and consider co-morbidities, motor and non-motor symptoms. In addition to the workflow and/process support, the system provides the organization tools for **risk management**, **standardization** and **data reports** that enable a better understanding, control and ways to manage the treating teams.

The care models are moving from **reactive** to **proactive approach**. The treatment of PD patient involves both approaches since it combines monitoring of motor symptoms, alerting and responding, as well as overall preventive assessment and on-going treatment in a proactive way. The REMPARK system receives data transmitted from the sensors. For each patient, there is a defined range of normal and abnormal values in each monitored parameter. Any deviation from normal event creates an alert on the system and consequently there is a reactive intervention. The **alerts** can be set not only for a single event but also for a set of events overtimes. For example, a fall requires immediate response therefore a single event will create an alert. However, for the neurologist it is important to know if the drug treatment is effective; therefore, parameters such as number of dyskinesia events or FoG events in the last week will be important and indicative for the treatment. In this sense, it is possible to set an alert whenever a set of accumulative events in a defined period occurs and to present it to the neurologist.

In addition, each patient has a **personalized treatment plan** defined by the doctor that is monitored by the nurse in the centre, who has a concrete list of **tasks** to handle. The tasks will be generated automatically from the patient's treatment plan or by the doctor/coordinator of the team. This way the patient is also being treated in a proactive way by the nurse or other care team members. The nurse can also see the condition of the patient in the last week at a glance or enter his record. The proactive treatment is driven by the treatment plan that is observed and followed by all care team members. This allows an **integrated care**, team discussion and coordination in providing comprehensive care. The platform has, also, the ability to communicate between care team through messaging tool or video call.

For each patient, participating in the REMPARK pilots, according to the patient's evaluation and clinical history, medications and definition of normal ranges for every measurement, the medical team can add and update the treatment plan according to changes in condition of the patient. A change in the treatment plan can be as a result of changes in the patient's health state or patient's environment. Some items create a task for a specific medical team member according to his profession while all team members can be updated in the status and given treatment.

The strength of the system also relies on the ability to obtain, analyse and **integrate data from different sources**: sensors, patient reported data and EHR. Wearable sensors provide data monitoring of clinical data about PD patients motor condition but also behavioral data such as activity levels, type of activities and social activity. The development of these technologies that monitor different data types to yield additional information about wide range of parameters and activities including behavioral, mental and clinical data that are transmitted in order to assist the treatment process, opens new opportunities for care providers but also requires integration of care models.

The data can be further analyzed for risk assessment of stratification for timely intervention. In these cases, data entered by the patient regarding his condition is pre-defined by the care provider and used as part of the treatment and for further assessment of the patient's condition [10, 11]. However, today the data collected for chronic conditions and specifically **patient reported data** in different care programs is partially used in care and are usually not treated by physicians as clinical data in the **EHR** as part of the clinical process. (see Figure 8.1 for a shot view of a screen showing patients' data. Consider that data are not real).

The challenge of healthcare system will be to use these data providing additional information to that existing in the EHR. The REMPARK system provides a tool for implementation of this approach by using the integrated care platform to combine heterogeneous data from EHR, **sensors**, **analytical tools** and patients' self-reported data for **comprehensive care** of multidisciplinary teams. Furthermore, the data are continuously updated from the different sources and do not require manual update enabling all stakeholders to continue using their organizational systems without duplication of work and bringing added value to the treatment process using intelligent tools on top of their existing system. The interaction between the patient and the healthcare professionals will be enabled through the DMS platform as a result of alerts and/or treatment plan, since the nurse will be able to call the patient or message him.

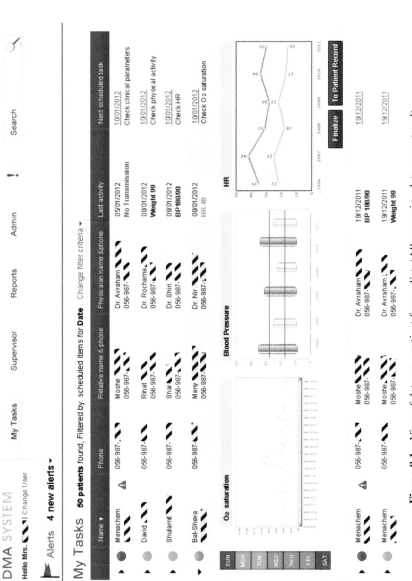

Figure 8.1   View of data on a patient from a list. (All appearing data are not real).

All correspondence is registered on the system for follow-up and **risk management**. One of the useful tools presented in REMPARK system is the **clinical protocols** tool. A clinical protocol is a document with the aim of guiding decisions and criteria regarding diagnosis, management and treatment of PD. It is a set of steps allowing the medical team to make decisions according to the collected information. It is actually a set of medical instructions for every situation according to the patient's medical state.

These protocols can be easily updated or changed from time to time by the care provider according to the medical considerations or according to organizational procedures. The protocols can guide the medical team how to treat the patient and co-morbidities. For this purpose, the clinical team of REMPARK project developed a set of clinical protocols based on the best practices for the treatment of PD patients. In addition, protocol for other chronic diseases or geriatric conditions exist in the system and enable the medical team to identify the clinical situation. The protocols are adapted to the user (e.g. healthcare organization) and can be generated for doctors, nurses or/and caregivers. By using the protocols the medical team will provide a standardized treatment that is in accordance with clinical guidelines (see Figure 8.2).

The protocol is loaded automatically in case of treating a task or an alert. The clinical protocol automatically guides the medical team user to the next

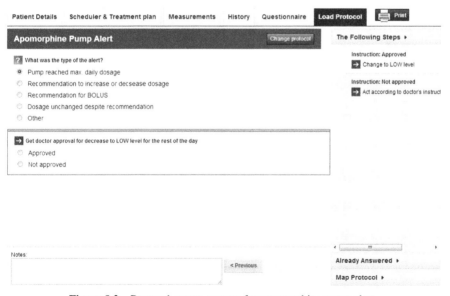

Figure 8.2 Protocol screen capture for apomorphine pump alert.

point or stage according to the data, until reaching the final decision or action. Using the clinical protocol the medical team will be able to send instructions to the auditory cueing system as described in the external interfaces or make a decision about any intervention or treatment using future tools that would be connected to the system.

In order to have an overall understanding about the patient's condition the neurologist interviews the patient about their symptoms, frequencies and daily condition. The patient is usually asked to fill a diary; however, the experience shows that these diaries are only partially filled in and the information provided to the neurologist is partial. One of the ways the REMPARK system overcomes this problem relies on the data obtained and analysed by the sensors and presented to the neurologist on the DMS platform. This tool is complemented under some conditions by the use of a questionnaire.

The DMS platform provides a tool of **questionnaires** that serves as "**patient reported data sensors**". The questionnaires can be filled by the patient, relative or by a nurse calling the patient in cases the patient is unable to fill the questionnaire. In addition, some of these questionnaires can also be responded through the smartphone, which are also received by the DMS. In either case, the system presents the patient the questionnaire that has to be filled according to the treatment plan. For each questionnaire, there is a logic to calculate a score. The medical team and the patient are able to see all history of previously answered questionnaires.

The questionnaire tool is also combined with the sensors data and can ask the patient additional questions when the sensor data alerts. The information can be viewed along time to show a trend even if no alert was raised. This tool was designed to collect the patient's reported data and use it with clinical logic to create alerts and response by the medical team. It integrates the data with the clinical data in an innovative way that presents the physician the information they need and simplify the analysis of the patient's reported data for use in clinical practice. Figure 8.3 shows, as an example, a part of the questionnaires section.

As healthcare moves towards proactive treatment there is a need for changes in relationship between the care provider and the patient. The proactive concept that involve the patients in self-care requires **patient empowerment**. The patient becomes an active partner in care and makes his own decisions about his health and disease. In order to do so the patient has to be educated about his disease, aware of the clinical options and learn how to cope with the disease. Patients who participate in their care process, cope better with their disease, live healthier lifestyle, have higher quality

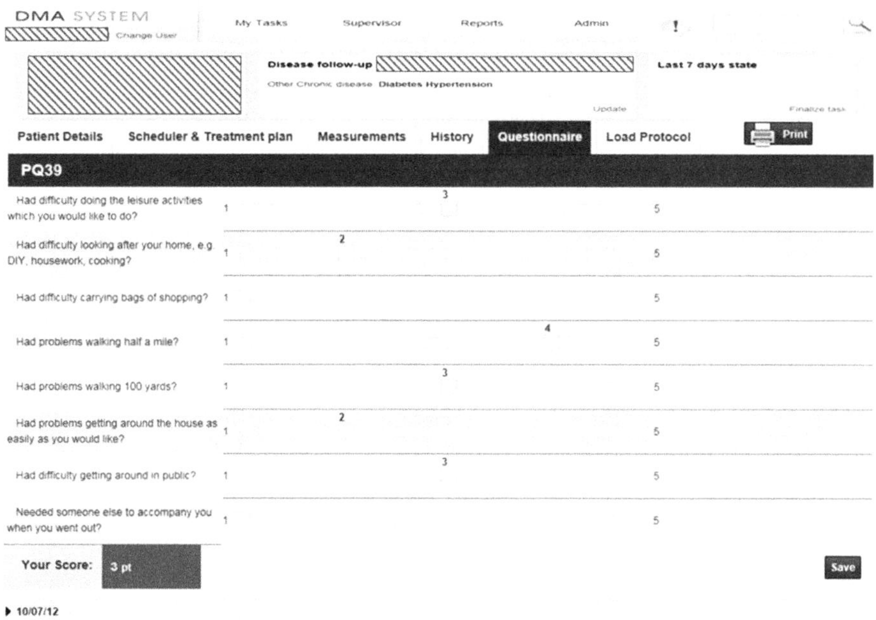

Figure 8.3 Questionnaire's section detail.

of life and better clinical outcomes [9]. For these reasons, the REMPARK system adopted this approach and developed a web portal and smartphone interface for the patient.

Electronic correspondence between the patient and the medical team is bi-directional. The correspondence can be triggered from the patient (using the patient web site or the smartphone) or from the medical team or any DMS user. The patient interface is designed to enable the patient receiving all the information about his condition. They can view all sensor's data, questionnaires and correspondence and perform actions required for the treatment such as filling questionnaires and for correspondence with their carers (healthcare providers). The patient is able to add activities into the treatment plan. The patient's web interface is used, among the above, to communicate with the patient either by global message to every person registered to the REMPARK system or messaging a specific patient/medical team member. This way the message is personalized to a patient or can be addressed to a group of patients with the same clinical condition or need.

An emphasis was made on **patient empowerment** and **education**. Educational materials are sent to the patient according to their condition, need, and disease progression such as articles regarding PD, new treatments and medications or any links to external internet pages and support groups. Figure 8.4 is showing how it looks the patient's interface presenting the personal treatment plan.

The REMPARK system provides additional organizational tools for management of the care team. The DMS includes a **shift management** module that manage the tasks and users that are at each shift. The shift manager can see the workload and type of activities done in the centre for efficient management of the care process. They can also receive reports about activities done on the system.

The above description of the REMPARK system provides an insight not only to the workflow application but also to the **care approach of REMPARK project** for PD patients. The system supports the care approach and enables the care provider to effectively manage the delivery of integrated care. The **uniqueness of REMPARK system** is that it includes a wide

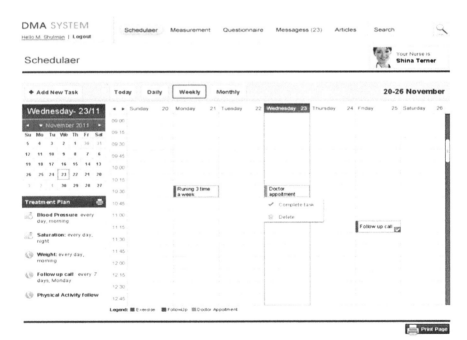

Figure 8.4   Treatment Plan. A patient's interface view.

range of functionalities for delivering comprehensive customizable integrated care by multidisciplinary team, but it also enables the treating physician to maintain a supervision of the patient throughout the journey; from patient disease diagnosis, through treatment plan definition, customizing specific protocols.

Moving from reactive to proactive treatment, involving all care providers, requires a change in interaction and mode of delivery the care under the physician coordination. Here, the care is delivered under his supervision enabling to effectively leverage other providers to pursue a consistent treatment approach and objective. It establishes a way for patient's personalized treatment. The application interfaces with the EHR, and the physician receives on-going alerts, updates and patient's data. The care team operates as a supporting envelope to the physician, and maintain constant touch with the patients according to the treatment plan, protocols and alerts received from the system.

8.3 DMS Functional Organization

For the already presented functionality, the DMS platform will communicate with the mobile gateway in both directions through the REMPARK server. This way, two types of information are distinguished:

- Incoming data: contains the patient's parameters as they have been measured by the sensors or/and patient processed information from the Rule Engine. Furthermore, the DMS will get information from the Rule Engine through the REMPARK server too.
- Outgoing data: contains updates for the patient's treatment according to the patient's pump or any other command from the auditory cueing system. These data are sent from the DMS.

The system architecture allows that all the data are stored at the REMPARK server, so the interaction between the different parts of the system will be with the REMPARK server. Figure 8.5 shows the main role that the REMPARK server has in the communications process. This is the only interaction point between the REMPARK's server to the Medical Application – DMS system.

In general terms, there are two types of communications: telemonitoring of motor symptoms and another related to patient management issues. These flows were implemented by sending and receiving different services between the server and the DMS system. In case of the tele-monitoring of motor

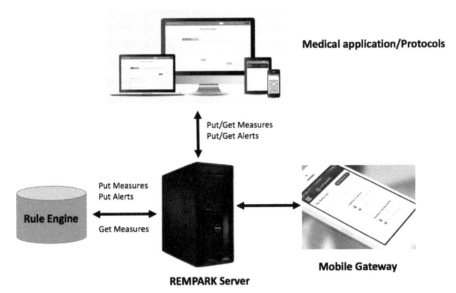

Figure 8.5   The server acting as a communication hub interfacing the DMA System.

symptoms, the DMS system will receive measures and alerts immediately from the REMPARK server. On the other hand, regarding the patient's management communication, the DMS system will send events regarding the medication, questionnaires and clinical agenda of the patient.

Figure 8.6 presents the implemented architecture for the Internet interface, including a web server and a data base server (can be a physical or virtual server).

8.4 Advantages Using the Disease Management System (DMS)

8.4.1 Advantages for the Clinical Team

Nowadays, Healthcare delivery systems face problems such as the prevailing inequities in access to care, resulting from services with availability problems in different geographic areas, socioeconomic and cultural disparities. In addition, the system is usually fragmented and a comprehensive patient-centred care is almost impossible to be delivered with existing systems and care models. Therefore, even if services exist, it is very complicated for the clinician to provide the services to many patients that are not located in the geographical environment and even for these patients, there is a need of coordination between service provider's.

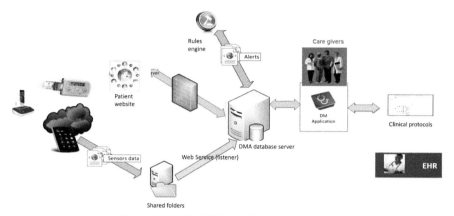

Figure 8.6 The Web interface architecture.

The main health dimensions are described in the SHARE document [14] and they can be listed as:

- Physical health measured by self-reports on general health, functional limitations, ADLs, IADLs, diseases symptoms, and health behaviour
- Healthcare use
- Cognitive function test, literacy, numeracy, memory, verbal fluency
- Mental health
- Physical performance measures

The inefficiencies, limited coordination and integration of complex healthcare systems further complicate the treatment and uneven adherence of patients to evidence-based medicine becomes a major barrier in treatment efficacy.

The care of PD patients is complex and requires a coordinated care team. It is possible to consider two approaches for delivering disease management programs: integrated and non-integrated care. Disease management programs based on the integrated care model with primary care have shown the most valuable and effective care and promising results. An integrated approach requires much deeper involvement from providers, and patients throughout the process and many health plans are starting to use in-house nurses with data analytics [12].

New models of integrated health management address member needs across the organization, offering value and transparency in all member interactions with the health plan. In order to provide such services there is a need in predictive modelling health risk assessment, utilization management and coordination of care. The care delivery model includes capturing and analysing patient data from dispersed systems, involving and engaging all stakeholders, especially primary care providers that are typically left out. An optimal identification and stratification of patient population is, also, derived from data integration and physician engagement challenges.

The REMPARK system supports the referred care approach and enables the clinical team to effectively manage the patient's disease. The advantage is that different healthcare professionals can use the same platform without changing their own system. The data are updated and the workflow is maintained. It serves for Information and knowledge sharing between all caregivers of the patient. The REMPARK system provides the care team a way to overcome the barriers of fragmented care service, overcome organizational and inter-organizational limitations and focus on patient's clinical needs for diagnosis to most advanced conditions, better management of care and replacement of some of the home visits by remote evaluations as well as better management of visit scheduling according to need instead of the periodic visits usually done.

At the PD management level, the system provides information that does not exist today enabling the neurologist to have better assessment of the patient's condition and treatment effectiveness with online information. The pilot studies done in REMPARK project provided clinical outcomes for the motor symptom detection. Further studies should be done for assessment of overall clinical outcomes of the new treatment approach on the PD patients. However, from previous experience with other chronic disease patients treated in Maccabi healthcare services, in Israel, using this approach, the following significant clinical outcomes can be found:

- Significant improvements in the patient's well-being measured by Qol.
- Patients maintained a healthy lifestyle: diet, physical activity...
- The mental condition of these patients (reported depression) dropped from 24% before to 15% following the intervention using DMS.
- Patients reported that they have better ability to cope with their disease.

At the level of physicians there was a high satisfaction from the service which they found very helpful in the treatment process since the system supports Healthcare professionals in their routine work.

8.4.2 Advantages for the Patients

Daily communication with caregivers, better understanding of the disease, education and awareness about the disease progression, symptoms and possible treatments are crucial for PD patients and their ability to cope with the disease. At early stages of the disease, there is a need to support the patient with educational materials and follow-ups, and at more advances stages there exists a difficulty of the patient to arrive to the clinic as often as needed, to adjust the medical treatment, provide additional treatments (physiotherapy and cognitive treatment, for instance) and perform a correct on-going assessment and intervention of the disease.

The REMPARK system monitors and provides care for PD patients considering their overall clinical condition including co-morbidities. The patients are treated by their doctor and do not have to move visiting different doctors and care providers for the treatment of different conditions. The care coordination is done within the program and provides them a clear treatment plan and one contact point. The system aims to enable the patients preserve/enhance of their physical and mental QoL, performance scores, compliance rates and satisfaction. It also enables to empower the patients and support them and their family in the care process.

In summary, the REMPARK system brings an innovative solution to the patients by involving them in the care process. They are actively participating in the assessment through the questionnaires and are able to view and follow their clinical condition. The neurologist can make a remote visit and assessment of the motor condition (administering the UPDRS scale, as demonstrated during the REMPARK project pilot in Israel) through video conference and save the need to go to the hospital, especially for patients that are living in rural areas or have motor symptoms and disabilities at advanced stages of the disease.

From the cumulated experiences with chronic disease patients adopting the DMS system approach at Maccabi in Tel-Aviv (Israel), the presented system clearly helps the patient in all the mentioned aspects. Patients seem less worried regarding their condition; feel they have the information for the treatment, for drug regimen and for self-management. It also helps to preserve/enhance their physical and mental QoL (as measured by PHQ9).

The treatment increases their performance scores and satisfaction, decrease reported depression and increase healthy behaviour. The effects were shown also on the whole treatment perception, self-management and treatment burden healthy lifestyle was assessed by the frequency of self-care activities of the patient.

- Concerning the diet, physical activity and drug regime, there was a significant improvement due to the intervention (satisfaction with the service is very high and is rated 6 in a scale of 1–7).

- Use of a DMS based service results in a better empowerment of the patient and caregivers support of the care process. An increase of compliance and adherence to treatment is also detected.
- Remote- UPDRS was performed in PD patients as part of the pilot.

8.4.3 Advantages for the Organization

It is very well-known that the prevalence of chronic diseases is increasing around the world, and 50–80% of all global health spending is related to chronic diseases. Patients suffering from Parkinson's Disease are part of this population, since in most cases, PD patients suffer from more than one chronic disease.

The care of PD patients is complex and the correct treatment should be handled jointly by General Practitioners (GPs), and the specialized health service (hospitals or clinics): specialists (geriatrics and neurologist), general physicians, nurses, physiotherapists, nutritionists, occupational therapists, social workers, that may be affiliated to different organizations. Rehabilitation and self-management support should be handled in cooperation with GPs [13].

Disease management programs focus on patient monitoring and intervention. These shift healthcare expenses to less invasive and expensive care, thus, disease management programs are meant to strive to achieve two seemingly conflicting goals:

- Improving patient outcomes
- Optimizing resource utilization

In parallel to the increase of the prevalence in chronic diseases, there is a decrease in number of healthcare professionals. Since the early 1980s there is a decrease in number of physicians per capita from $\sim$3 to $\sim$2 doctors per 1000 people in average in the OECD countries. The number of nurses per 1000 people decreased from 7.5 in the mid-1980s to 5.5 in 2004 in average in the OECD countries. This trend increases the workload on physicians and nurses as well as on other healthcare professionals which limit the system in providing all the services needed.

The healthcare system structure is fragmented and in order to treat PD and other chronic patients, it is needed to receive services from several stakeholders. There is a limited accessibility and availability and lack of treatment uniformity in different areas according to healthcare professionals and service availability. There is a continuous rise in national healthcare costs and as a result, the economic stress is rising.

Nowadays, the healthcare systems face huge challenges. At the level of the care process there is a need for capturing and analysing patient data from dispersed systems, involving and engaging all the related stakeholders, especially primary care providers that are typically left out. Optimal identification and stratification of patient population derives also from data integration and physician engagement challenges. Service should be equal to all patients regardless of their living area (rural or large cities) and involve care teams and case managers. At the organizational level, there is a need to establish the process/workflow, the clinical guidelines and protocols for remote treatment, risk management and cost control.

As it has been already explained, the integrated care workflow application considered in REMPARK includes built-in alerts, clinical protocols, guidelines and treatment plans. The application interfaces with remote monitoring devices and can be integrated with EHR and other medical systems. The advantage is that the system does not change the routine work of the healthcare professional and support them in their work and decisions, enabling knowledge sharing, online assessment, intervention and follow-up. Using this approach on PD patients, in REMPARK project, it was received a high level of satisfaction from both patients and doctors, however it was detected a real need to drive towards the implementation of the service and to study the success measurements as it has been already done in other chronic disease patients populations.

The system is composed of several modules that provide a number of tools and functionalities to the organization:

- Risk management.
- Control of large teams providing services 24H/7D.
- Management of a large groups of patients that will enable scaling up to large population management.
- Treating heterogeneous data coming from different monitoring device, including patient reporting data.
- Ability to treat different conditions from well-being, clinical cognitive and mental conditions.
- Keep more than one care organization in the loop by sharing data and knowledge.
- The support of several care models/workflow and therefore the minimization of the number of systems that the organization has to use and implement.

8.5 Conclusions, Discussion and Vision

As the healthcare services changes and moves from hospitals to community care, there is a need to address the need of coordination between the services provided by specialists in the hospital, such as neurologists, and healthcare providers in the community and to create a system that will enable continuity of care, global coordination and integrated care. Furthermore, the treatment of actual complex patients includes also social care organizations, municipalities and sometimes private care providers, thus making their care quite complex, fragmented and with difficulties for sharing relevant information.

On the other hand, the interaction and relationship between patients and care providers is changing and is moving towards collaboration schemes and the active involvement of the patient in his care process, including self-management. This absolutely requires new tools and a different approach for patient empowerment, guidance and timely response by the care provider.

The REMPARK project developed and experienced a system that aims to address the challenges of the future model of care for the PD community. The different levels of sensors, data analytics, decision support tools, integrated care platform and actuators as well as tools for the patient, demonstrate the process and care pathway for treating PD patients and chronic patients at home.

The success of the system relies on its ability to adapt to different organizational and national systems as demonstrated in 3 pilot sites (Spain, Ireland and Italy) in which the main care of PD is provided by hospitals and in Israel in which the main treatment is provided in the community. The reliability of the system also at the level of data analysis provides a powerful tool when combined with clinical and organizational guidelines and thus pave the way to treat large populations with different and changing conditions in a personalized way.

The main barrier in implementing such systems is the need of integration with organizational systems, which is a slow and partial process. Therefore, the main challenge of healthcare system is to implement such systems at large scale and to use them as part of the existing workflows, sharing information with other organizations, patient and family and to develop methodologies for interactive treatment.

Healthcare organizations will use these technologies by changing the model of care, sharing information and advancing the research and development of systems that will bring additional information and knowledge to that already exist, in order to find the best way to use it in practice to improve care quality. This will also change the clinical practice and patient involvement in self-care and decision making.

References

[1] Andlin-Sobocki P, Jnsson B, Wittchen HU, Olesen J (2005). Cost of disorders of the brain in Europe. Eur J Neurol 12:1–27.

[2] Armstrong RA (2008). Visual signs and symptoms of Parkinson's disease. Clin Exp Optom 91(2):129–138.

[3] Davie CA (2008). A review of Parkinson's disease. Br Med Bull 86(1):109–127.

[4] Hou JGG, Lai EC (2007). Non-motor symptoms of Parkinson's disease. Int J Gerontol 1(2):53–64.

[5] Jankovic J (2008). Parkinson's disease: clinical features and diagnosis. J Neurol Neurosurg Psychiatry 79(4):368–376.

[6] Korczyn AD (2008). Parkinson's disease. In: E. in Chief: Kris Heggen-hougen (ed) International encyclopedia of public health, Academic Press, Oxford, pp. 10–17.

[7] Samii A, Nutt JG, Ransom BR (2004). Parkinson's disease. Lancet 63(9423):1783–1793.

[8] Sian J, Gerlach M, Youdim MBH, Riederer P (1999). Parkinson's disease: a major hypokinetic basal ganglia disorder. J Neural Transm 106:443–476.

[9] Hadas Lewy. Future challenges for implementation in healthcare services. Healthcare Technology Letters. Special issue: Wearable Healthcare Technology, Vol. 2 Issue 1, pp. 2–5, 2015.

[10] Lewy H. Kaufman G. The Maccabi multidisciplinary center for disease management in "Recipes for sustainable healthcare" sponsored by abbvie, Philips and EUPHA. Brussels, Biblioteque Solvay Brussels, May 2013.

[11] Yee K., Yeung A., Robertson T., et al.: 'A networked system for self-management of drug therapy and wellness'. Wireless Health '10 October 2010, San Diego, CA, USA.

[12] Matheson, M., et al.: 'Realizing the promise of Disease Management – Payer Trends and Opportunities in the United States', The Boston Consulting group, 2006.

[13] Joseph F. Coughlin et al.: 'Old Age, New Technology, and Future Innovations in Disease Management and Home Health Care', Home Health Care Management & Practice, Vol. 18, 3, 2006.

[14] The Survey of Health, Ageing and Retirement in Europe – http://www.share-project.org/

9

REMPARK System Assessment: Main Results

Albert Sama[1], J. Manuel Moreno[1], Carlos Pérez[1], Joan Cabestany[1], Angels Bayés[2] and Jordi Rovira[3]

[1]Universitat Politècnica de Catalunya – UPC, CETpD – Technical Research Centre for Dependency Care and Autonomous Living, Vilanova i la Geltrú (Barcelona), Spain
[2]Centro Médico Teknon – Grupo Hospitalario Quirón, Parkinson Unit, Barcelona, Spain
[3]Telefónica I+D, Spain

9.1 Introduction

In the previous chapters, the REMPARK system has been presented and discussed, along with its usefulness. According to the initial project specifications and some important conclusions obtained from the work carried out during the project, the main characteristics and requirements of the integrating parts and REMPARK sub-systems were initially defined and refined afterwards.

The present chapter will first describe the final system and, then, its overall final assessment. This way, first, a description of the communication flows established in the REMPARK platform (type of information, dataflow and security issues) is given. Secondly, an overview of the final necessary platform, ready to be used in the piloting part of the project is provided.

The assessment process was done through a pilot in which the patients tested the system in ambulatory conditions and during their daily living activities. More than 40 PD patients participated from four countries (Spain, Italy, Israel and Ireland).

179

9.2 Description of Main Communication Flows

As it has been described in previous chapters, the effectivity of REMPARK system is based on a couple of interaction loops that define specific communication data flows. This way, communication between the different sub-systems of REMPARK is a crucial aspect and the correctness and security of the information flows must be guaranteed. In order to understand and specify the characteristics of the communication flows, it must be considered that the REMPARK system has two differentiated functional parts:

- The first one is related to the **monitoring of PD motor symptoms**, including: movement sensors, actuator, smartphone, the REMPARK server, the Rule Engine, the Disease Management System (DMS) and a patient's web interface.
- The second functional part is related to the **monitoring of non-motor PD symptoms**. This monitoring strategy is mainly based on the use of questionnaires and also includes a simple agenda system to interact with the patient. This part of the system requires the smartphone, the REMPARK server and the DMS to exchange information.

Thus, two different information flows are distinguished in the REMPARK system. This differentiation will be used to specify the system according to either 1) the communication flow due to the monitoring of motor symptoms and the application of the actuation action or 2) the communication flow due to non-motor symptoms.

- The monitoring process of the motor symptoms is mainly done in the immediate loop, around the patient. The mobile phone is acting as the gateway for the communication of the related sub-systems with the REMPARK server and the different channels and data type, according to the definition presented in the next section, are shown in the Figure 9.1:
- The non-motor symptoms mode embraces the communication flows related to patient management issues, e.g. agenda, treatment, questionnaires, etc. Figure 9.2 presents the REMPARK system parts participating in non-motor symptoms monitoring and their communication flow.

9.2.1 Type of Transmitted Information

This section details the general specifications of the different information types sent and received through the communication channels indicated in Figures 9.1 and 9.2. Four different types of information are considered:

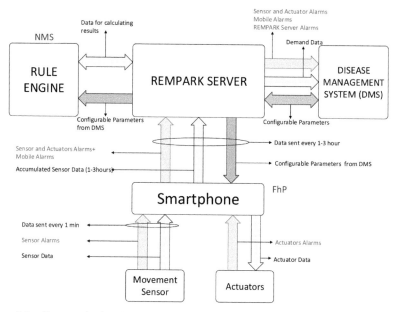

Figure 9.1 Communication flow of motor symptoms detection and actuation in REMPARK system.

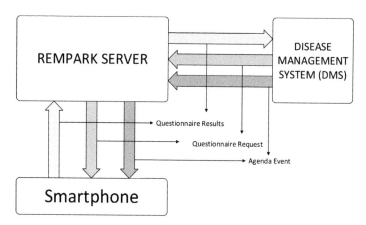

Figure 9.2 Communication flow of non-motor symptoms management and actuation in REMPARK system.

(I) Alarms, (II) Data sent within the closed loop of the REMPARK monitoring system, (III) Configuration parameters in the smartphone and, finally, (IV) Agendas and questionnaires:

- **Alarm specifications**
  An alarm is defined as an event detected by the REMPARK system that must be urgently notified to an end-user. For instance, a fall-detection event is the only alarm provided by the REMPARK wearable system. Other alarms can be defined within the DMS as a combination of one or more rules applied to the stored information; for instance, a value from a questionnaire lower than a specific threshold.

  In the communication flow, as a general rule, alarms within the REMPARK system will be handled as states. This means, for instance, that when the sensor sends a data packet it will also send the state of all the alarms, a value 1 meaning an activated alarm and a value 0 meaning a deactivated alarm. There should be 3 stages for clinical alarms: red, yellow and green. Therefore, when a low battery level is detected its corresponding alarm will be set to 1, and its value will not become 0 until an optimum charging level is detected.

- **Data sharing between the REMPARK wearable system and the DMS**
  The sensor sends data every minute to the smartphone. The smartphone will store the data sent by the movement sensor during a specified time frame, which usually consists in few hours. Once this time frame is reached, the smartphone establishes a connection with the REMPARK server to upload all the data stored since the last connection. The connection between the smartphone and the server will take place every 1–3 hours, except when the data sent by the sensor contains an activated alarm or when an alarm is generated at the smartphone. In this case, an immediate connection with the server is established.

- **Configurable parameters**
  All configurable parameters must be stored in the REMPARK server. These parameters are used by the machine learning algorithms and the ACS cueing system, which enable their personalisation to each patient. They are defined in the DMS and their values are transmitted to the smartphone, in which they are applied to the sensor's output provided every minute. These parameters are, for instance, the patient's walking rhythm, the threshold used to detect bradykinesia or the freezing index level necessary to consider an episode of FOG.

- **Questionnaires and agendas**
  The data associated with the monitoring of non-motor symptoms can be grouped in three broad categories: the request for a specific questionnaire, the generation of an event in the agenda and the completion of a questionnaire. All the questionnaires, including the TAP tests, are stored

locally in the smartphone and in the DMS. The request for a specific questionnaire consists in asking for a specific questionnaire number and a date and time. The results of the questionnaires are sent in an encoded REMPARK format. As an option, the questionnaires could be filled in 3 ways:

1. Using the smartphone
2. Through the Patient's web application
3. By the nurse asking the patient and completing, directly, through the DMS platform

The data from the questionnaires will be sent to the REMPARK server from the smartphone or the DMS respectively, they will be analysed by the Rule Engine and alerts will be then sent to the DMS. The smartphone will send all data filled by the patient and will be presented to the patient also on the Web application. An agenda event is generated by the DMS and sent to the smartphone through the REMPARK server. Even if the event is stored in the server for record purposes, this information will not be accessible by the rest of the subsystems (smartphone or DMS) once the event is transferred to the smartphone. All the data related to the questionnaires and the agendas must be stored in the REMPARK server.

9.2.2 Security Aspects

Security aspects related to the communication are very important in the application of the REMPARK system. The system can be split into two different domains (the BAN and the platform/server), and different considerations should be made possible. In the BAN, security issues are related to the Bluetooth communication between the different sub-systems. In the platform, security aspects come in the communication between the smartphone and the server and the rest of the entities of that domain, the Rule Engine and the DMS. The most relevant security aspects are summarized as follows:

- **BAN:** All the different devices forming the BAN, i.e. the sensor, the smartphone and the cueing system, exchange data using Bluetooth 2.1. Traditionally, using Bluetooth devices was associated to use pin codes for each device in order to establish communication between devices. This method required a manual intervention and security is not very high. REMPARK uses Bluetooth v2.1, which uses SSP. This security approach uses public key cryptography, which gives a lot of advantages. It just works without manual intervention; however, a device may prompt the user to confirm the pairing process. During the pairing time, if the devices have a screen, numeric comparison is used, which informs the user that a device is trying to connect. The number shown in the screen must be the same as the one in the device. From Bluetooth v2.1 encryption is required and the encryption key is regularly refreshed.

- **Platform/server:** The server is interfaced by three different elements, the Rule Engine, the DMS and the smartphone. Communication with the server fulfils the European Data Protection Directive (Directive 95/46/EC), which embraces the different national laws for data protection.

9.2.3 Final Deployment and Implementation of the REMPARK Platform

With the already presented specifications and characteristics, the REMPARK platform was implemented for its use during the last piloting phase of the project. According to the pilot specification, the system was conceived to be implemented following a distributed approach, and this is how it was tested during an early pre-pilot phase. Many efforts were devoted to ensure the synchronisation, data transmission, processing and representation of the information considering different scenarios like a sudden lack of communication, for instance.

After this early phase, the integration of the Rule Engine into the REMPARK server was decided, obtaining in this way a more efficient management of the communication bandwidth and a mitigation of transmission delays.

The REMPARK system worked steady, secure and with good performance to carry out the scheduled pilots. The Figure 9.3 shows the **final REMPARK functional view** (it was deployed in June 2014 as starting point for the pilots).

In the figure, the Rule Engine is already presented as a part integrated in the REMPARK server and directly accessing to the information generated by the patient's sensor (measures, alerts). All this information is transmitted by patient's smartphone in a secure way using a Transport Layer Security (TLS) channel.

The Rule Engine, then, analyses this information and creates new postprocessed measures, which are stored and forwarded to the DMS system in a secure way using a TLS channel. Eventually, the DMS shows all the gathered information to the clinical professionals as well as providing a decision support system that follows the defined clinical protocols. Furthermore, clinical professionals can set up treatments and appointments that are forwarded to patients' mobile phone.

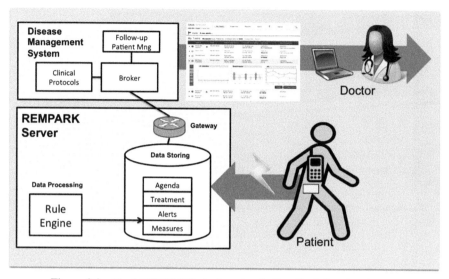

Figure 9.3   Final REMPARK platform (ready for pilots) functional view.

In order to guarantee the execution of the pilot, it was decided to implement four different and equivalent environments. As REMPARK pilot experience must be distributed among four different countries (Spain, Ireland, Italy and Israel), it was entailed the deployment of four different servers, one isolated server per pilot site and the same replica for the DMS system.

In order to guarantee patient's data privacy, pseudonyms and ID's instead of personal information were transmitted. Moreover, to maximise the security of the pilots, external calls to each environment were forwarded through a proxy to an isolated Telefónica's private network which can only be accessed from inside (Telefónica was the responsible partner for the REMPARK platform and communications). The Figure 9.4 shows how it looks like this set up.

In total, four environments were set up assuring this level of security. Furthermore, any external communication with the server, either from patient's mobile phone or the DMS system goes through a TLS channel and each message includes specific credentials according to the type of user/system sending the message (i.e. patient, DMS). Moreover, other security mechanisms have been implemented to assure that there are warning alerts in cases where there are communication problems between the different devices (i.e. alert when the smartphone loses communication with the platform, alert if sensor or actuator connections are lost, presence of a fall detection). Figure 9.5 shows

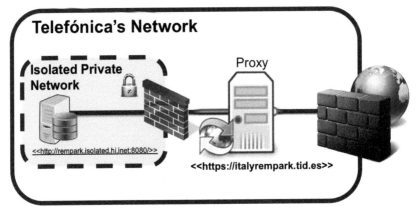

Figure 9.4   REMPARK server in isolated network.

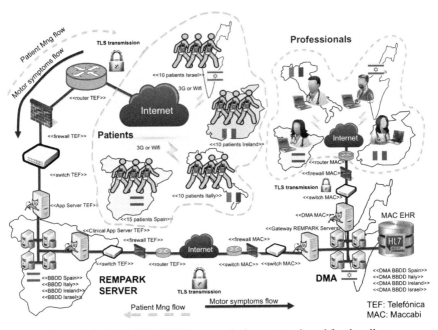

Figure 9.5   Final REMPARK system deployment and used for the pilots.

the final set-up implemented and used for REMPARK's pilots. The secure connections between the REMPARK server and other external sub-systems, like the DMS and the smartphones used by the patients are indicated. As a proof of concept, the DMS was integrated with the information systems used by Maccabi in Israel through HL7 protocol.

9.3 Pilot for the REMPARK System Assessment: Description

This section introduces the details and the execution of the distributed pilot, executed in REMPARK. The main objective of this pilot was the global assessment of the system in terms of ON and OFF detection. As a previous activity, before starting the main pilot, an early pre-pilot was organized with very few volunteers in order to test the stability and the validity of the overall system.

In this section, the experimental protocol is described first, followed by some conclusions after the early trial (pre-pilot experience). Finally, the analysis of the REMPARK system during the pilots is fully provided, with the main conclusions regarding the obtained results.

9.3.1 Definition of the Pilots' Objectives and Eligibility Criteria

The objectives of the pilot activity were divided into primary and secondary. The complete list to be covered is following:

Primary objectives

- To study the performance and reliability of the REMPARK system under real conditions on a significant group of patients.
- To study the validity of the ON/OFF motor phase detection by the system on a significant group of patients.

Secondary objectives

- To study the functionality of cognitive tests administered through the smartphone in detecting ON/OFF states on a pilot group of patients.
- To check the operation of an auditory cueing system, activated by the output of an algorithm, in improving the gait on a pilot group of patients.
- To assess both usability and user's satisfaction referring to the REM-PARK system under real daily living conditions on a pilot group of patients.
- To study the validity of the fall detection algorithm of the REMPARK's system on a pilot group of patients.

- To verify the safety of the REMPARK in individuals with Parkinson's disease on a pilot group of patients.

Population and eligibility criteria

The reference population is that formed by a pilot group of patients with moderate to severe Parkinson's Disease, presenting ON/OFF phases, FOG or dyskinesia (Hoehn and Yahr greater than 2 in ON phases and lower than 5 in OFF phases). Candidates found to be eligible to participate in the study immediately were provided with the Patient Information Sheet and the Informed Consent Form. The Principal Investigator or a co-investigator met the subjects and explained the study purpose, procedures, possible risks and benefits and subject responsibilities to the potential participants. The subjects had the opportunity to evaluate these documents in detail and were allowed to ask the investigators any question regarding the study. The subject's willingness and ability to meet the follow-up requirements were determined.

Patients fulfilling the following criteria were candidates to be included in the study:

- Clinical diagnosis of Idiopathic Parkinson's Disease according to the UK Parkinson's Disease Society Brain Bank [1].
- Disease in moderate-severe phase, with a Hoehn and Yahr greater than 2 in ON phases and lower than 5 in OFF phases, or with a scale greater or equal to 2.5 (in ON state) with motor fluctuations comprising bradykinesia, FOG and/or dyskinesia [2].
- Able to walk unaided in OFF state.
- Age between 50 and 80 years.
- Sufficient literacy capacity to answer questionnaires.
- Willing to participate in the study (in writing-sign informed consent) and wanting to co-operate in all its parts, accepting the performance regulations and procedures provided by the researchers. The patients' family members and responsible doctors should fulfil, also, this inclusion criteria.

Patients fulfilling the following criteria were excluded from the study:

- Other health problems that hamper physical activity and gait: rheumato-logic, neuromuscular, respiratory, cardiologic problems other neurolog-ical disorders- (Post stroke, Polio), mental disorders, significant pain.
- Alcohol and/or drug abuse.

- Being treated with duodopa or apomorphine pump, or deep brain stimulator.
- Patients who are participating in another clinical trial.
- Dementia according to clinical criteria -DSM-IV-TR [3].
- Unable to fully understand the potential risks and benefits of the study and give informed consent.
- Unable to recognize the ON and OFF fluctuations, after proper training.
- Subjects who are unable or unwilling to cooperate with study procedures (for example- unwilling to carry the sensors during the study hours).

All participants in the pilot presented motor fluctuations, according with the inclusion criteria, and at least 25% of the total of the sample had FOG problems. Family members and responsible doctors of the participants were included in the study and they were allowed to consult patient's health data through the web information system.

9.3.2 Design of the Study

The REMPARK pilot was designed as a longitudinal pilot study in which each participant was using the REMPARK system under real conditions (ambulatory conditions) during 4 consecutive days. An additional pre-trial day (day 0) is necessary for customizing the system to the user, and a day 1, to train the user in the system's use. This pre-trial day was scheduled at least 1 week prior to the trial start. The organization of the study design is shown in Figure 9.6.

In the pre-trial day (day 0), the thresholds for the bradykinesia detection algorithm, the FOG detection algorithm and the TAP test parameters were adjusted for each individual participant. For doing so, three short specific experiments were conducted: walk in OFF, walk in ON, FOG provocation test and TAP test. During the trial days (day 1–day 4), patients were wearing and using the REMPARK system under real conditions. The system was recording

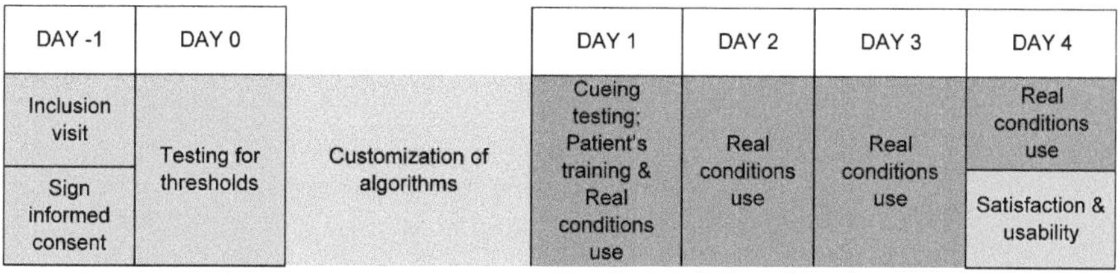

DAY -1	DAY 0		DAY 1	DAY 2	DAY 3	DAY 4
Inclusion visit	Testing for thresholds	Customization of algorithms	Cueing testing; Patient's training & Real conditions use	Real conditions use	Real conditions use	Real conditions use
Sign informed consent						Satisfaction & usability

Figure 9.6 Timing of the methodological study design.

motor and non-motor symptoms, which were also video recorded by standard means (video cameras or smartphones).

The pilot study started after getting the approval of the corresponding Ethics Committees and the competent authorities in the four countries (i.e., in Spain the Spanish Agency for Drugs and Medical Devices, AEMPS gave the corresponding permission). The following data and records from each participant were obtained and integrated into the pilot database:

- **Personal identification data:** name, address, phone number.
- **Socio-demographic variables:** date of birth, gender, educational level, cohabitation.
- **Clinical assessment questionnaires:**
  - **Parkinson's diagnosis:** according Brain Bank criteria by Hughes et al. [1].
  - **Parkinson's severity:** measured by Hoehn and Yahr scale [2].
  - **Questionnaire assessment of motor fluctuations, FOGs, and dyskinesia.**
  - **Basal health status:** list of chronic conditions, list of current medications.
  - **ON/OFF current phase:** The current motor phase was established by using three different instruments. 1) Patient's report: patients who were previously trained for recognizing their ON/OFF phases 2) report whether they are in ON or OFF phase every 60 minutes, or 3) when they detect a change.
  - **Unified Parkinson's Disease Rating Scale (UPDRS)** [4], which was administered by a trained researcher to participants several times during the pilot.
  - **Video recording** and posterior analysis by an expert was used for confirming the motor phase of the patient.
  - **TAP-test:** This test was designed to objectively assess changes in movement, such as slowing and rigidity which are effects commonly related to the OFF state. In this test the subject is requested to repeatedly press a button on the touch screen of the smartphone representing a white arrow in order to fill up a vessel as soon as possible according to ten predefined levels (each target level is marked by a green bar). The ten levels represent the number of button presses needed for the level to be fulfilled. The subject has to press the button by using his forefinger of the dominant hand (i.e., right hand).

- **Cognitive status:** Assessed by Folstein Mini-Mental [5] and applying the criteria of DSM-IV TR dementia.
- **Freezing of Gait (FOG) episodes and severity**, measured by two different standard methods:
  - Video recording and subsequent analysis by a trained observer, was used for identifying the section of the sensor signal's corresponding to FOG episodes.
  - Freezing of Gait Questionnaire (FOG-Q) [6]: This instrument was used for establishing the basal (over the last 4 days week) severity of freezing of gait in the participants, and also its severity during the 4 days of the REMPARK's system use.
- **Gait quality:**
  - Video recorded timed Up & Go test [7]: This instrument is a timed test of standing and walking. It is a gait-speed test used to assess a person's mobility and requires both static and dynamic balance.
- **Non-motor symptoms:**
  - Non-Motor Symptom Questionnaire: NMSQuest [8].
- **Usability:** The usability instruments were used to assess separately the usability of each device, and also the whole system.
  - System Usability Scale (SUS) [9].
  - Quebec User Evaluation of Satisfaction with Assistive Technologies (QUEST) [10].

The REMPARK system for these pilots included all the already described sub-systems:

- Sensor sub-system (located in the waist). μSD cards were used on day 0 to collect the raw data from the sensor in order to perform the user particularization. The remaining days, the sensor shared the results of the algorithms with the smartphone once per minute.
- Smartphone: The smartphone used in the experiments is the Samsung Galaxy Nexus. The specific software running in the smartphone was developed within the framework of the REMPARK project.
- Headset for auditory cueing: The headset chosen for the REMPARK experiments was the Samsung HM3500 Bluetooth earphone.
- REMPARK Platform (according the description in precedent text).
- Web Server – Disease Management System (application for managing the patient's health by the medical staff).

9.3.3 Pre-Pilot Conclusions

As it has been mentioned, a preliminary pre-pilot activity was scheduled with the participation of a limited number of patients, and with the objective of checking the stability of the overall REMPARK system to be used. It was also intended to evaluate its performance in terms of correct detection of the motor state of few PD patients in ambulatory conditions.

During the pre-pilot, regarding the system stability, it must be noted that the data gathered by the REMPARK sensor were correctly sent through the different communication layers so that no information loss due to communication errors were detected. Although some packet losses were observed during the experiment, they were due to the fact that the user didn't remember to have the smartphone always in coverage range of the sensor. In this respect, it was concluded that, during the user training session, emphasis to keep both devices as close as possible during the whole duration of the experiments had to be done.

It was also found with these pre-pilot tests that the performance of the REMPARK system when detecting the motor state of a patient is in line with the original specifications, even if the gold standard available to validate the results is not always accurate enough. The existing issue is the fact that the only available gold standard is the self-reporting of the patients using a diary. Figure 9.7 presents the diary filled by one of the participants in the pre-pilot experience, during the first day of experimentation (day 1). It can be observed that the patient, as it is commonly done in motor states self-reporting, writes down the motor state for each hour of the day.

Reporting the motor state in this manner presents the following two issues:

- The time in which the motor state is reported has been shown to be inaccurate in some cases. More specifically, it is unclear when exactly changes among motor states occur. In contrast, the information obtained by REMPARK system is known with seconds' accuracy.

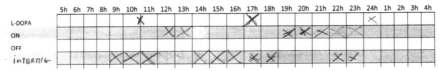

Figure 9.7 Motor states section of the diary filled by one patient during the first day of experimentation. Third row corresponds to the 'intermediate state'.

- <u>Some PD patients are not able to recognize their motor state</u>. This risk could be mitigated in the recruitment phase, since neurologists only should select those patients able to recognize ON and OFF states.

In order to evaluate the ability of the REMPARK system for the detection of ON and OFF motor states, a set of three main performance measures are used: specificity, sensitivity and correlation with a gold standard. These measures can be severely affected by the two inaccuracy issues presented above:

- Concerning specificity and sensitivity measurements: These measurements may be strongly affected by the described issues, since the REMPARK system provides an estimation of the motor state once every 10 minutes. First, given a patient who constantly switches among different motor states, a big difference between the time assumed in self-reporting and the time provided by the REMPARK system could lead to very low specificity and sensitivity values. Second, if the patient does not correctly recognize the motor state, a similar effect would be obtained.
- Concerning the correlation between REMPARK system output and patient diary (used as gold-standard), it must be pointed out that these measurements would be affected similarly to the specificity and sensitivity ones.

Two possible options appear when comparing the self-reported motor state of a patient and the REMPARK system output:

- It is possible to consider the correct REMPARK system output as the output that is closest to the time of the self-reported motor state by the patient. Results for this option were presented as '*exact time*'.
- It is possible to consider the correct REMPARK system output as the mode (most frequent value) among the outputs obtained within the one-hour period that corresponds to the self-reported motor state. Results for this option were presented as '*mode*'.

Finally, it should be noted that patients can report to be in ON, OFF or intermediate state (see Figure 9.7). However, REMPARK system only provides ON and OFF predictions. In consequence, those periods in which intermediate state was reported will not be included in the performance analysis.

Table 9.1 shows results for two patients, who participated in the pre-pilot. No valid conclusions can be extracted from this pre-pilot phase, but the identification of the discussed issues and the technical validation of the system. In the results of the first patient, it can be observed that only 8 hours were analysed, given that most of the reported states correspond to the 'intermediate state'.

Table 9.1   Pre-pilot results on motor state detection

Patient	REMPARK System Output Selection	Specificity	Sensitivity	VPP	VPN	Correlation	Number of Hours Analysed
1	Mode	0.6	0.3	0.33	0.6	–0.07	8
	Exact time	1	1	1	1	1	8
2	Mode	0.91	0.71	0.71	0.90	0.62	29
	Exact time	0.73	1	1	0.54	0.7	29

9.4 REMPARK Pilots' Execution and Obtained Results

Pilots with REMPARK system were conducted in accordance with the protocol presented above. It must be remembered that the main objectives of the study were to analyse the performance and reliability of the REMPARK system under real conditions, and to study the validity of the ON/OFF detection by the system. A list of secondary objectives was also focused (see Section 9.3.1).

The recruitment was carried out following a convenience sampling among patients assisted in the different centres participating in the study, following the methodology described in the above Section 9.3.1. Fifty-four (54) patients were initially contacted, 44 of them met inclusion criteria and agreed to participate. Three of the included patient did not complete the study days. One of them discontinued participation voluntarily. Two patients were removed from the study by the researchers, the first of them due to lack of adherence to the study protocol, the second patient was removed due to a health condition, which required hospitalization, apparently not related to study devices or procedures. Finally, 41 patients completed all study days and evaluations.

Table 9.2 shows recruitment and follow-up data by participant entity.

Table 9.2   Recruitment and follow-up data by participant entity

Medical Partner	Contacted Patients	Included Patients	Lost-Dropout Patients	Completed Protocol
NUI Galway (Ireland)	7	7	1	6
TEKNON (Spain)	18	15	0	15
FSL (Italy)	16	10	0	10
MACCABI (Israel)	13	12	2	10
Total	54	44	3	41

Concerning the socio-demographic and health data of the participants, twenty-eight participant patients were men (68.3%) and 13 were women (31,7%). The average age of participants was 71.3 years (SD 7.3; range 56–84).

All the participants had been diagnosed of Parkinson's Disease according to the UK Parkinson's Disease Society Brain Bank [1]. The average time from diagnosis was 11.3 years (range 2–26). As per inclusion criteria, all participants had a Hoehn and Yahr scale equal or greater than 2.5 [2], being median stage of the sample 3 (range 2.5–4). The motor section of the Unified Parkinson's Disease Rating Scale (UPDRS) was in average 28.2 (SD 1.7) [3], when the patients were in OFF phase and 12.2 (SD 1.5) when the patients were in ON phase. The median of the Freezing of Gait (FOG) questionnaire of all participants was 13 (IQR 6) [4], this scale explores gait disturbances of the patients and ranges from 0 through 24, being the higher scores related to worse gait disturbances.

Participants with dementia and acute medical conditions were not included in the study. Mini-Mental State Examination (MMSE) [5] results of the participants are shown in Table 9.3, along with other medical chronic conditions of participants.

Table 9.3 Mental and chronic disease condition of the participants

Condition	Number of Participants (n)	%
High blood pressure	15	36,6
Heart conditions	3	7,3
Arthritis, osteoarthritis or rheumatic conditions	11	26,8
Back ache	13	31,7
Asthma or COPD	1	2,4
Diabetes	10	24,4
Urinary incontinence	14	35
High cholesterol	12	29,3
Depression	9	22,5
Anxiety disorder	3	7,3
Stroke, cerebral embolism, cerebral infarct or cerebral bleeding in the past	0	0
Cancer (malignant tumours)	5	12,2
Osteoporosis	2	5
Thyroid disease	1	2,4
Cognition (MMSE)	29 (Median)	5 (IQR)

9.4.1 Performance of the System

The assessment of the system performance was done considering and analysing the data received and stored in the REMPARK server.

One of the main problems that was detected in the system and that was widely repeated in many users is related to the communication structure of the BAN and the usage received from patients. REMPARK wearable system is composed of a smartphone, a wireless headset and the waist-worn movement sensor. Bluetooth is employed among them to communicate. More concretely, the movement sensor and smartphone communicate once per minute (every 60 seconds). In consequence, given the short-range reached by this type of communication, whenever patients placed the sensor and smartphone at a certain distance (e.g. in different rooms) communication could be lost. Some measures would not be received by the smartphone which, then, would not be received by the server. These periods of lost information were mainly addressed and analysed as an objective measurement of the system performance.

In order to distinguish them from those in which the sensor is turned off, it was used the first packet after the communication between the smartphone and the sensor is established, which contains specific values that make this situation recognisable.

The performance analysis process of the REMPARK system was organized according the following procedure. First, presenting a summary of the most relevant data sent and stored in the server during the pilots, distinguishing them according to the different sources and types of information. Then, data received in each day of the pilot were analysed in order to evaluate the performance of the system communications. In ideal conditions, in every minute during which the patient wore the system, the corresponding set of samples should be stored in the server.

Table 9.4 shows just a summary example of the kind of data stored in the REMPARK server. This table includes the source of the information (in this example, only data corresponding to the sensor and the smartphone are shown), a description of the variable (between parentheses, the internal code assigned to the variable in the database is appearing) and the amount of data samples for each variable stored in the server.

Along the complete pilot execution, a total of 949 hours of raw data were stored in the system. This assumes that an average of 6.6 hours of raw data per user per day were collected. It is noted that this average number of monitoring hours is obtained on a minute basis, that is, the number of values in which the movement sensor sent information, meaning that missing values are not included.

Table 9.4 Example of data stored in the REMPARK server

Information Source	Information Variable	Quantity
Movement sensor	BRADYKINESIA – NUM STEPS (106)	56955
	BRADYKINESIA – BRADY MEAN (107)	56955
	BRADYKINESIA – BRADY STD (108)	56955
	DYSKINESIA – DISK PROBABILITY (110)	56955
	DYSKINESIA – DISK CONFIDENCE (111)	56955
	FOG – MAX_FI (112)	56955
	ACTIVITY – SMA (901)	56955
	ACTIVITY – CADENCE (902)	56955
Smartphone	BRADYKINESIA (201)	56955
	FREEZING OF GAIT (202)	56955
	DYSKINESIA (203)	56955
	TAP RESPONSE TIME MEAN (204)	99
	TAP RESPONSE TIME STD (205)	99
	LAST TAP RESPONSE TIME MEAN (206)	99
	LAST TAP RESPONSE TIME STD (207)	99
	TAP TEST RESULT DECISION (208)	99

In order to estimate the amount of missed data, it must be considered that if the sensor did not communicate with the smartphone during a given period of usage, missing packets are reflected with time intervals higher than 60 seconds. More concretely, it is considered that some packets have been missed when two consecutive movement-sensor measurements were received by the smartphone with more than 100 seconds of difference.

It was observed that some packets were lost in most patients (only in 5 patients there were not any missed packets). Moreover, it was calculated that the average number of missed packets was only of 6 minutes. Finally, the error rate was considered in terms of missed packets vs. total number of measurements received. According to this, it was observed that less than 4% of the packets were missed among all patients.

These missed short periods are not relevant since motor symptoms algorithms, require sensor measurements on a time interval of 60 minutes, allowing to have missing values in the analysed period, which means that these short-time missed packets do not influence the information provided to clinicians.

Given these considerations, Table 9.5 provides the performance analysis of REMPARK system related to a very few patients participating in the pilot (shown patients are randomly selected from the complete set and must be considered only as an illustrative example).

Table 9.5   Performance analysis of the REMPARK system (some examples)

Patient Number	ID	Number of Time Intervals of Missing Packets	Average Number of Missing Packets Per Time Interval	Number of Time Intervals > 10 min.
1	TEKNON 1	1	2,67	0
2	TEKNON 2	18	10,55	6
16	MACC 1	37	19,74	8
24	FSL 1	6	3,28	0
25	FSL 2	11	7,64	2
34	NUIG 1	21	20,11	10
Total pilot average	–	8,64	6,70	2,16

9.4.2 Validity of the ON-OFF Detection Algorithm

REMPARK system would be very useful enabling neurologists to adjust the medication regime based on detailed information of ON/OFF states fluctuations, which is provided in real-time by the wearable sensor subsystem (a waist-worn inertial sensor), with the concurrency of a smartphone. Final adjustment is then performed with the help of the specifically designed Disease Management System (DMS), which is a server-based service that allows neurologists, patients and caregivers to follow the disease evolution and communicate among them. This section presents the validation of the main objective of the REMPARK: the monitoring of ON/OFF motor fluctuations. The presented validation consists in the evaluation of the REMPARK system in detecting motor states in PD patients, who used the system over 4 consecutive days, as part of the described pilot.

9.4.2.1 The methodology

The system assesses, in real-time, objective ON/OFF motor states based on the inertial signals given by the waist-worn device, treated with different machine learning algorithms that provide the presence or absence of specific symptoms, from which PD motor states are then determined. In the following part, the ON/OFF detection methodology is described.

Accelerometer measurements obtained by the waist-sensor are analysed with two different algorithms: a bradykinetic-gait detector [11] and a dyskinesia (choreic-dyskinesia) detector [12]. The bradykinetic-gait detector is based on the patients' strides analyses, only after identifying that the patient is walking. Each stride is characterized by a measure of its fluidity: high fluidity measurements correspond to non-bradykinetic gait and vice-versa. On the other hand, the dyskinesia algorithm has two main outcomes: first, a probability value that represents the chances of having obtained the signals from a patient suffering dyskinesia and, second, a confidence value that represents the degree of certainty on such probability. The outputs of both algorithms on a period of T minutes are then processed.

The inferred motor state is ON when bradykinesia is not present in gait or when dyskinesia output is present. This is due to, on the one hand, a lack of symptomatic gait in ON states and, on the other hand, because of peak-dose dyskinesia detection, also is associated to ON states. OFF state is determined as the detection of bradykinesia and the absence of dyskinesia. Finally, the intermediate state is defined by an intermediate detection of the bradykinesia algorithm and lack of dyskinesia. A not evaluated (NE) state is provided in case of not detecting dyskinesia and if the patient did not walk in the period under analysis.

A final refinement is performed to the sequence of ON-OFF detections obtained from the evaluation of 10 consecutive 1-minute periods. Those periods determined as not NE are changed according to the following rules:

- If a NE period is found between two periods whose states are equal (either ON or OFF), that period is set to be the same state as adjacent periods.
- If adjacent periods of an empty period correspond to different motor states, an intermediate state is then inserted.

9.4.2.2 Validation of the ON-OFF diaries: available data

As it was introduced in above Section 9.3.3, the only possible and available gold-standard for the validation of the ON and OFF states detection algorithms are the patients' diaries with the self-reported states and all the inherent and already described inaccuracies. In order to stablish a methodology, two strategies were followed to ensure the validity of the gold-standard:

- First, the validity of motor state diaries was evaluated based on clinicians' expertise through UPDRS. In each scale that clinicians administered, it was annotated the state in which the patient perceived to be. As it has been described in many works devoted to study the factors involved in UPDRS, scores obtained in OFF state are higher than the scores obtained in ON states [13]. Thus, the validity of the gold-standard has been evaluated based on an objective measure consisting in the correlation between UPDRS scores and motor states provided by patients at the moment of administration. Motor state is represented with 2 = OFF, 1 = Intermediate and 0 = ON, so positive correlations are expected. It is noted that, in some cases, negative correlations have been obtained. In order to also avoid those situations in which there is not any relation between UPDRS and the annotated motor state (i.e. nearly-zero correlation), patients whose correlation was lower than 0.20 were not included in the study since their ON/OFF diaries were considered not reliable enough to be used as gold-standard.

- A second aspect taken into account is the timeframe during which annotations are valid. More concretely, in order to consider a motor state annotation to be valid, this motor state has to appear two consecutive times in the diary, being the time interval of 1 hour between both annotations its corresponding valid timeframe. In case the motor state changes between two consecutive annotations, both annotations are excluded from the analysis since the time in which the motor state changed cannot be established. Although this strategy reduces the number of annotations used from patients, it ensures that the annotations employed to validate the system are temporally reliable. Results obtained when this procedure is applied are presented under the name *Strict diary*. On the other hand, when this procedure is not applied, results obtained are reported as *Original diary*.

The complete validation of the ON/OFF detection algorithm requires the usage of the data stored in REMPARK server during the pilots, the ON/OFF diaries filled in by patients and the CRF's administered by clinicians. Finally used and useful data for these purposes were obtained from a total of 36 patients, after two drop-outs and the exclusion of 5 diaries due to incoherencies (15 patients administered by Teknon in Spain, 10 patients from Italy administered by FSL, 8 patients from Israel under the supervision of Maccabi and 3 patients from Ireland administered by NUIG).

9.4.2.3 Results of the ON/OFF state detection

Among the initial 36 patients, data from 3 of them were removed due to a low correlation between UPDRS scores and the movement state reported by the patient (correlations were –0.62, 0.06 and 0.2, respectively), as discussed in the previous section.

Table 9.6 presents the average results of REMPARK system in detecting motor states in the 33 patients under real-life conditions, being the system adjusted without using data from artificially induced OFF states. The average specificity and sensitivity achieved by the system in recognizing

Table 9.6 Average results according to the Original diary method and time-ensured method (Strict)

Validation Method	Average Specificity Per Patient	Average Sensitivity Per Patient	Average Number of Validated Hours Per Patient	Average Number of ON/OFF Monitoring Hours Per Patient
Original diary	82%	57%	9.73	81.25
Strict diary	89%	98%	6.42	81.25

ON/OFF motor states is 89% and 98%, respectively. These values dramatically decrease when the described time ensuring is removed: specificity is 82% and sensitivity falls to 57%.

In overall, the system generates an average of 35.3 hourly-based motor state detections per patient over the 4 days (11.8 hours in ON, 4.9 hours in OFF and 18.6 hours in an intermediate state). Thus, in average, for each of the four days, almost 9 hours of monitoring per day have been provided by the system.

Performance obtained by the system provides excellent results (sensitivity of 98% and specificity of 89%). Although the minimum sensitivity achieved is 75%, some specificity values are presented in the range of 50–70%. The lowest values are obtained due to a very low number of validated estimations for some of the participating patients. It must be noted that in some cases validation was not possible due to a lack of annotations, which is noted with NE in both Specificity and Sensitivity (main reason was because patients did not fill in the diary).

Regarding the method used to set the thresholds of the bradykinetic-gait measurement algorithm, the distribution-based approach reveals to suitably set the algorithm parameters, according to the excellent results obtained. The novelty of this approach relies on the fact that the approach does not re-quire any OFF-induced data. In contrast, other works such as [14] employ OFF-induced states, obtained by skipping medication intakes, to train machine learning algorithms. Following the method described here, patients would not be required to skip medication intakes to adjust the algorithms. Moreover, patients do not have to follow any specific set of scripted activities as in [14], either at home or in a lab, to particularize the detection method.

One of the main limitations of the system is that, in order to estimate the motor state, it requires the patient either to walk or to present choreic-dyskinesia. However, this is compensated by a reasonable high number of monitored hours with enough number of correct detections and a correct selection done for the kind of patients recruited (H&Y greater than 2 and lower than 5 in OFF state). On the other hand, one of the main advantages of the motor states detection offered by REMPARK system is in the fact that the evaluation method only requires patients to wear a single sensor.

9.4.3 Non-Motor Symptoms Descriptive Analysis

The REMPARK system was designed for the administration of questionnaires to the participants in the pilots [15], through the mobile phone interface. One of these questionnaires was a patient-based 30-points, used to determine the non-motor symptoms experienced by the patient during the past month. It had thirty items answered by the patient with a dichotomous answer: yes or no.

The following two questions were added to enquiry other non-motor symptoms:

- Difficulty to speak.
- Have you had unusually strong urges that are hard to control? Do you feel driven to do or think about something and find it hard to stop (Such gambling, cleaning, use the computer, obsessing about food or sex)?

In addition to the standard dichotomous answers (yes or no), the REMPARK system was given the option of rating each symptom by using an analogical 10-points scale (0–10).

All the 41 patients answered the non-motor symptoms questionnaire (NMSQ). Among them, 63.4% presented mild non-motor symptoms, 36.6% presented moderated non-motor symptoms and no patient had a result within the "severe" range of the questionnaire. The median of the questionnaire answers was 8 (over the 30 points) and IQR was 9.0.

Table 9.7 shows a random selection of answers (five of them) with the most frequent positive answer. In addition, the two new introduced questions are also shown.

9.4.4 Efficacy and Effectiveness of the Cueing System

The efficacy of the cueing system to improve walk was tested by comparing the timed UP and GO test [16] with and without the cueing system activated, according the protocol of the study established in REMPARK.

The average time performing the timed UP & GO (TUG) test with the cueing system activated was 25.8 seconds, and without the cueing system was 25.7 seconds. Thus, no significant or clinical difference was found.

Table 9.7 Non-motor symptoms selected answers sorted by frequency

Question Number	Question Text	Positive (n)	Positive (%)
8	A sense of urgency to pass urine makes you rush to the toilet	27	69,5
9	Getting up regularly at night to pass urine	27	69,5
5	Constipation (less than 3 bowel movements a week) or having to strain to pass a stool (faeces)	17	41,5
12	Problems remembering things that have happened recently or forgetting to do things	17	41,5
15	Difficulty concentrating or staying focused	17	41,5
Added	Difficult to speech	26	63,4
Added	Have you had unusually strong urges that are hard to control? Do you feel driven to do or think about something and find it hard to stop (Such gambling, cleaning, use the computer, obsessing about food or sex)?	2	4,9

Several sub-analyses selecting participants were performed with severe gait problems according their FOG questionnaire. Some improvement in the average time in completing the timed UP and GO test was found then, but it was still no significant from a statistical point of view. Results of these sub-analyses are shown in Table 9.8.

The effectiveness of the ACS cueing system to improve walking problems was tested by comparing the results of two administered FOG questionnaires (before and after the REMPARK system testing). At the beginning of the pilots, participants answered the FOG questionnaire [6], which measures walking related problems (higher values in the results of this questionnaire means more

Table 9.8 Results of the cueing systems for participants with severe gait problems

FOG Questionnaire Filter*	n	TUG with Cueing: Average Seconds	TUG without Cueing: Average Seconds	Mean Differences (seconds)	Comment
Questions 4 > 2	11	24,7	27,0	−2,4	Ns
Questions 5 & 6 > 2	9	26,9	29,5	−2,5	Ns
Questions 4 or 5 or 6 > 2	16	24,5	26,0	−1,5	Ns

*FOG questionnaire question 4 > 2: longest FOG episodes lasting > 10 seconds
*FOG questionnaire question 5 > 2: starting hesitation episodes lasting > 10 seconds
*FOG questionnaire question 6 > 2: turning hesitation episodes lasting > 10 seconds
Ns: no significant.

severe gait problems), regarding the last 4–5 days. After the pilot experiments, the participants answered again the same questionnaire, regarding the pilot days.

The mean of the FOG questionnaire before using the REMPARK system was 12.8 and after the use of REMPARK system was 12.3, not being the difference clinical or statistically significant. The results did not significantly change in the sub-group of patients with worse walking problems. Thus, we cannot conclude with the data obtained in the study that the cueing system of the REMPARK system helped the patients to reduce their walking problems in real conditions.

9.5 Health-Safety of the REMPARK System

In all health interventions, there is a risk that intervention entailing unexpected negative effects that counteract the possible benefits from it. It is important to notice that, according to the operative definition, and adverse event is "*any untoward medical occurrence, unintended disease or injury, or untoward clinical signs (including abnormal laboratory findings) in subjects, users or other persons, whether or not related to the investigational medical device*".

During the experimental time, 7 participants presented health "adverse events", one of them had a "serious adverse event" consisting in condition deterioration that led to hospitalization. The most frequent adverse events were limbs pain (three cases) and depressive symptoms (3 cases); cervical pain was also reported (1 case). All adverse events, including the serious adverse event, were considered as "*unrelated to the investigational device*".

In conclusion, we can't identify special health-risk associated with the use of the REMPARK system in this pilot, although the study timeline and the lack of comparison group makes difficult to detect any health-risk associated with the use of the system. Further studies on safety are warranted.

9.6 Usability and User Satisfaction of the REMPARK System

The usability and user satisfaction with the system were measured by using two standard instruments: the System Usability Scale (SUS) [17] and the Assistive Device sub-scale of the Quebec User Evaluation of Satisfaction with Assistive Technologies (QUEST) [18].

The SUS is a 10 items Likert scale in which the respondent indicates the degree of agreement or disagreement with the statement on a 5-point scale (Figure 9.8). To calculate the SUS score, first the score contributions from each item are summed. Each item's score contribution ranges from 0 to 4. For items 1, 3, 5, 7 and 9 the score contribution is the scale position minus 1. For items 2, 4, 6, 8 and 10, the contribution is 5 minus the scale position. Finally, the sum of the scores is multiplied by 2.5 to obtain the overall value of SUS. SUS scores have a range of 0 to 100.

Results over 50 are considered acceptable, and over 68–70 are considered good [19, 20]. The median SUS score of the REMPARK system in the pilots was 70 (IQR 25).

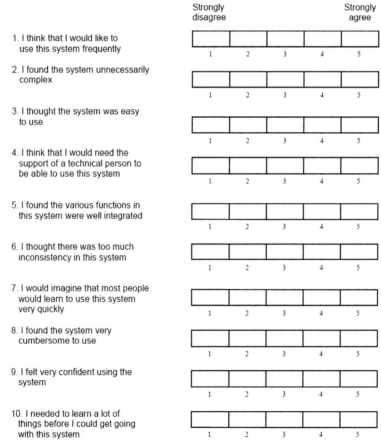

Figure 9.8 System Usability Scale.

The results can be considered good, according with standard interpretations of the scale. However, the results must be interpreted with caution, as the researchers administered themselves the SUS to the participants, who knew that the answers were not anonymous. This could have biased the results, because participants could have tried to please the researchers, offering a more positive vision of the usability than they would have given under anonymous circumstances.

Concerning the evaluation of user satisfaction, a subset of the QUEST questionnaire was used. The QUEST scale as it is described in [18] contains a number of items related to device characteristics and several questions related to the service characteristics. All the answers must be done according the following scale:

1. not satisfied at all
2. not very satisfied
3. more or less satisfied
4. quite satisfied
5. very satisfied

For evaluating REMPARK system, the 6 following items of the device subscale were used, together with a general question (number 7):

1. The dimensions (size, height, length, width) of your assistive device?
2. The weight of your assistive device?
3. The ease in adjusting (fixing, fastening) the parts of your assistive device?
4. How safe and secure your assistive device is?
5. The ease in using your assistive device?
6. The comfort of your assistive device?
7. What is your overall satisfaction with the assistive device?

Obtained results are shown in Table 9.9. The results can be considered good, being comfort the worst rated item and safety the best rated item. Overall satisfaction is good, being not satisfied 5% of the participants, and quite satisfied or very satisfied 75% of the users.

9.7 Summary and Conclusions

Forty-one (41) Parkinson Disease patients, under real conditions, participated in the test of the REMPARK system. Overall results show that the system has a good performance and reliability, with few data loss (4% of data), which does not impact the main information provided by the system, as it comes from several data slots, grouped and presented in 60 minutes' periods.

Table 9.9   Results for the QUEST questionnaire

	Not Satisfied At All		Not Very Satisfied		More or Less Satisfied		Quite Satisfied		Very Satisfied	
	n	%	n	%	n	%	n	%	n	%
Dimensions	0	0	3	7,3	10	24,4	13	32	15	36,6
Weight	0	0	2	4,9	6	14,6	14	34	19	46,3
Ease in adjusting	1	2,4	1	2,4	15	36,6	11	27	13	31,7
Safe	0	0	0	0	7	17,1	22	54	12	29,3
Ease in using	0	0	2	4,9	8	19,5	22	54	9	22
Comfort	3	7,3	2	4,9	11	26,3	18	44	7	17,1
Overall satisfaction	0	0	2	4,9	8	19,5	19	46	12	29,3

The validity of the REMPARK system to detect motor fluctuations is very good. When all possible errors of the gold-standard timeline (the diary completed by the user) are excluded, by using the so called "*strict method*", the sensitivity reaches 89% and specificity reaches 98%. These seems very good results, given the fact that they have been achieved in real life conditions, which are very demanding; in real life, there are many situations which could cause false positive or false negative detections. We consider this an important outcome, as it opens the door to a possible clinical use of the REMPARK system in the future.

The study has also demonstrated how difficult it is for patient to fill their diaries. In this respect, the REMPARK system constitutes a promising candidate to evaluate the correlation between the medication intake and the motor state in real life conditions. This can also constitute an invaluable help as a quantitative assessment instrument in clinical trials.

Additionally, during the REMPARK pilots, users could answer non-motor symptoms questionnaires using the mobile-phone interface. We did not measure the validity of this way of administration of questionnaires, but we really didn't expect a decrease in validity of the questionnaires when administered in this electronically way. This, again, suggested that the REMPARK system could be successfully used by clinicians, to remotely gather non-motor symptoms information from their patients.

Although the usefulness of the cueing system to improve gait in PD patients have been previously demonstrated by other groups, we failed to show any benefit of the REMPARK system in improving gait, during this specific pilot. No benefit was noticed when performing the timed Up and Go test under controlled conditions (efficacy), though a tendency to improvement was shown in the subgroup of patients with severe gait problems. These results did not reach statistical significance, possibly due to the small sample size, which pose problems of statistical power. In addition, the FOG questionnaire did not improve after using the system during the testing days (effectiveness). We think these results are inconclusive, as they were secondary outcomes of the pilot, which was not specifically designed to address these issues. If REMPARK system can improve gait problems or not, should be tested in longer experiments, specifically designed for measuring the effectiveness of the ACS sub-system.

The REMPARK system appeared to be safe, as no health adverse events were noticed, which could be related to the system. However, this result has to be interpreted with caution, as the small sample size and short observation period, pose again problems of power to detected adverse events. Safety of the system should be tested in additional research studies.

Finally, REMPARK system appears to be usable and participants seem to be satisfied with the system. Although this is a very encouraging conclusion, a possible bias in the usability results has to be taken into account, as the participants' answers were not anonymous and the researchers where not blind to their answers.

Overall, we consider that the REMPARK system piloting was successful, demonstrating good performance and validity to detect motor fluctuations, and good usability characteristics. The system is prepared for further testing in its way to commercialization and medical use.

References

[1] A. J. Hughes, S. E. Daniel, L. Kilford, and a J. Lees, "Accuracy of clinical diagnosis of idiopathic Parkinson's disease: a clinico-pathological study of 100 cases.," J. Neurol. Neurosurg. Psychiatry, vol. 55, no. 3, pp. 181–184, Mar. 1992.

[2] M. M. Hoehn and M. D. Yahr, "Parkinsonism: onset, progression, and mortality. 1967.," Neurology, vol. 57, no. 10 Suppl 3, pp. S11–26, Nov. 2001.

[3] American Psychiatric Association. (2000). Diagnostic and statistical manual of mental disorders (4th ed., text rev.). Washington, DC: Author.

[4] T. P. Fahn Stanley, Jenner Peter, Marsden C. David, Recent developments in Parkinson's disease. Florham Park, NJ: Macmillan, 1987, pp. 153–163.

[5] M. F. Folstein, S. E. Folstein, and P. R. McHugh, "Mini-mental state," J. Psychiatr. Res., vol. 12, no. 3, pp. 189–198, Nov. 1975.

[6] N. Giladi, J. Tal, T. Azulay, O. Rascol, D. J. Brooks, E. Melamed, W. Oertel, W. H. Poewe, F. Stocchi, and E. Tolosa, "Validation of the freezing of gait questionnaire in patients with Parkinson's disease.," Mov. Disord., vol. 24, pp. 655–661, 2009.

[7] D. Podsiadlo and S. Richardson, "The timed 'Up & Go': a test of basic functional mobility for frail elderly persons.," J. Am. Geriatr. Soc., vol. 39, pp. 142–148, 1991.

[8] K. R. Chaudhuri, P. Martinez-Martin, A. H. V. Schapira, F. Stocchi, K. Sethi, P. Odin, R. G. Brown, W. Koller, P. Barone, G. MacPhee, L. Kelly, M. Rabey, D. MacMahon, S. Thomas, W. Ondo, D. Rye, A. Forbes, S. Tluk, V. Dhawan, A. Bowron, A. J. Williams, and C. W. Olanow, "International multicenter pilot study of the first comprehensive self-completed nonmotor symptoms questionnaire for Parkinson's disease: the NMSQuest study.," Mov. Disord., vol. 21, no. 7, pp. 916–23, Jul. 2006.

[9] J. Brooke, "SUS – A quick and dirty usability scale," in Usability Evaluation in Industry, 1996, pp. 189–194.

[10] Demers L., Weiss-Lambrou R., Ska B. Development of the Quebec User Evaluation of Satisfaction with assistive Technology (QUEST). Assist Technol. 1996; 8(1):3–13.

[11] Rodríguez-Molinero, A., Samà, A., Pérez-Martínez, D. A., López, C. P., et al. Validation of a Portable Device for Mapping Motor and Gait Disturbances in Parkinson's Disease. JMIR mHealth and uHealth, 3(1), e9, 2015.

[12] Samà, A., Pérez-López, C., Romagosa, J., Rodríguez-Martín, D., Català, A., Cabestany, J., Pérez-Martínez, D.A., & Rodríguez-Molinero, A. Dyskinesia and motor state detection in Parkinson's Disease patients with a single movement sensor. In Engineering in Medicine and Biology Society (EMBC), 2012 Annual International Conference of the IEEE (pp. 1194–1197). IEEE. August 2012, San Diego (CA, USA).

[13] Vassar, S. D., Bordelon, Y. M., Hays, R. D., Diaz, N., Rausch, R., Mao, C., & Vickrey, B. G. Confirmatory factor analysis of the motor unified Parkinson's disease rating scale. Parkinson's Disease, 2012.

[14] Tzallas, A. T., Tsipouras, M. G., Rigas, G., Tsalikakis, D. G., Karvounis, E. C., Chondrogiorgi, M., Psomadellis F., Cancela J., Pastorino M., Arredondo Waldmeyer M. T., Konitsiotis S. and Fotiadis, D. I. PER-FORM: A System for Monitoring, Assessment and Management of Patients with Parkinson's Disease. Sensors, 14(11), 21329–21357, 2014.

[15] Romenets S. R., Wolfson C., Galatas C., Pelletier A., Altman R., Wadup L., Postuma R. B. Validation of the non-motor symptoms questionnaire (NMS-Quest). Parkinsonism RelatDisord. 2012 Jan; 18(1): 54–8. doi: 10.1016/j.parkreldis.2011.08.013. Epub 2011 Sep 14.

[16] D. Podsiadlo and S. Richardson, "The timed 'Up & Go': a test of basic functional mobility for frail elderly persons.," J. Am. Geriatr. Soc., vol. 39, pp. 142–148, 1991.

[17] J. Brooke, "SUS – A quick and dirty usability scale," in Usability Evaluation in Industry, 1996, pp. 189–194.

[18] Demers L., Weiss-Lambrou R., Ska B. Development of the Quebec User Evaluation of Satisfaction with assistive Technology (QUEST). Assist Technol. 1996; 8(1):3–13.

[19] Bangor A., Kortum P., Miller J. Determining What Individual SUS Scores Mean: Adding an Adjective Rating Scale. Journal of Usability Studies, 4(3) pp. 114–123.

[20] System Usability Scale (SUS): http://www.usability.gov/how-to-and-tools/methods/system-usability-scale.html Accessed May 27th 2015.

10

Epilogue and Some Conclusions

Roberta Annicchiarico[1], Angels Bayés[2] and Joan Cabestany[3]

[1]IRCCS Fondazione Santa Lucia (FSL), Rome, Italy
[2]Centro Médico Teknon – Grupo Hospitalario Quirón, Parkinson Unit, Barcelona, Spain
[3]Universitat Politècnica de Catalunya – UPC, CETpD – Technical Research Centre for Dependency Care and Autonomous Living, Vilanova i la Geltrú (Barcelona), Spain

10.1 Summary of PD Symptoms and the Influence on QoL

As it was extensively presented along Chapters 1 and 2, Parkinson's Disease (PD) is a major, chronic, non-communicable disease and the second most frequent neurodegenerative disorder worldwide. There is currently no cure for PD, but treatments are available to help relieve the symptoms and maintain individual's Quality of Life (QoL) at least for the first years.

It has been already debated the impact on the QoL of the Parkinson's Disease due to an enormous number of motor and non-motor symptoms: bradykinesia, rigidity, tremor, postural instability, reduced gait speed, freezing of gait, sleep disturbances, depression, psychosis, autonomic and gastroin-testinal dysfunction as well as dementia. As already explained, the majority of patients will develop an increasing number of more complex symptoms over time.

The treatment in the early stages of the disease, focused on the use of levodopa in pills, is very effective. Nevertheless, different problems related with the treatment may start to appear depending on the advance of the disease. Thus, it might be the case of motor complications: motor fluctuations such as the wearing-off phenomenon, involuntary movements known as dyskinesia, abnormal cramps and postures of the extremities and trunk known as dystonia, and a variety of complex fluctuations in other motor and non-motor functions. In these cases, the correct adjustment of the therapy is crucial for avoiding to decrease the QoL of the patient. The motor symptoms are especially responsible of falls and gait impairments and negatively impact on QoL by reducing the ability to perform many activities of daily living. They are the major causes of institutionalisation and to lose the capability to live independently. Daily tasks at home: self-care, food preparation, hygiene, become difficult, as do many activities outside the home (shopping, visiting friends/family, leisure activities …).

The global management of this disease has to be based on a simultaneous treatment and consideration of different symptoms that usually are treated in different by different specialists. Unfortunately, many times no integration between data is made and there is a lack of information about the overall condition of the patient and, furthermore, there is not enough communication between the healthcare professionals treating the patient.

REMPARK system, discussed along the present document is a good example of the recent advances in technologies for people with neurological diseases, focused on the development and validation of tools, techniques and overall solutions for the effective management of PD. The system exhibits the combination of two technologies that promise to radically change the daily management of PD: the so-called wearables technology (WT) and the machine-learning based algorithmic approach. These technologies can provide objective, high frequency, sensitive and continuous data on motor and non-motor phenomena in PD. Thanks to the technology, it is possible to solve one on the major existing problem: the data acquisition over a continuous time period. An important aspect of a WT system is that it allows remote monitoring of symptoms with its obvious potential advantage for patients and health economics.

10.2 Existing Barriers of PD Management

In order to preserve their quality of life and allow them to live independently for longer while experiencing their burdens, PD patients fully resort on tailored treatments that can address the symptoms as they appear or prevent/delay the onset of other symptoms and co-morbidities. Emerging systems, like REMPARK, try to improve the efficacy of disease management and treatments in the current clinical practice that presents the following major obstacles:

- **Barrier 1: Lack of accuracy and completeness when reporting about own symptoms.** Due to the cognitive impairments, distress or the evasive nature of some of the symptoms caused by PD, the patients often find difficulties or lack sufficient ability to provide reliable/consistent clinically relevant information about the symptoms they experience in order to optimise the treatment. In particular, often the patients are not aware of the onset of dyskinesia and sometimes it is even difficult for them to distinguish between ON/OFF periods. However, these are key information items for the doctor to adjust the treatments.

- **Barrier 2: Missing information about the PD symptoms and signs of disease progression at clinical level.** The current available means to report and monitor the symptoms are modest as compared to the huge challenge posed by the variety of PD symptoms and their fluctuations. The patient's visits and self-reporting may not throw reliable or complete evidence for the physician to cope with the entire picture and overall phenomena surrounding their patient's day-to-day. Most of the evidence used builds on reporting provided by the same patients and they often lack the ability to undertake this task.
- **Barrier 3: Compromised self-care and adherence to treatments.** Treatment regimens (medications, times, doses) and adherence to treatment are crucial for a correct PD management and the QoL of the patients. PD patients resort to prescribed regimes, but this seemingly simple commitment may represent a non-trivial feat, since patients must add on top of the overall burden the challenge of self-care, which is often difficult to achieve due to the many impairments and distresses linked to the disease. Cognitive deficits such as attention, communication, memory, and executive functions; depression and impulsive behaviours play a key role in the common lack of adherence and self-efficacy in co-management of the disease.
- **Barrier 4: Symptoms recognition in time to better administrate the medication dose.** Another related barrier is the capability of the professional to properly assess the number of OFF hours the patient has experienced to judge, based on that information, the therapeutic effect of the administered therapy, since it is based on daily or patient recall. Thus, the practitioner has difficulties for adjusting the continuous dose and control the administration of extra doses, mainly when an infusion pump therapy is used.
- **Barrier 5: Usability from the patient's point of view.** Some patients with Parkinson's have OFF phases so severe that they cannot even self-administrate extra doses. Patients with severe OFFs, which have no caregivers who can perform this task for them, often cannot choose the treatment with continuous infusion pumps.

10.3 The Role of the REMPARK System in the Context

REMPARK system was proposed and has been developed as a personal health system for the remote and autonomous management of PD. It is composed of wearable sensors for the detection and measurement of motor symptoms, a high level analytical layer and decision support tool, and a care platform (DMS) for professional care providers, patients and family. It enables the real-time, continuous and quantitative identification of the patient's motor symptoms in ambulatory conditions. The novelty and added value of REMPARK relies on using just one inertial sensor located and fixed to the patient's waist and a mobile interface to constantly read and send to the remote server a wealth of key relevant motor symptoms related to PD.

After symptoms' identification, the system is also able to provide some degree of actuation, like gait guidance mainly with an auditory cueing system, automatic fall detection and alarms, and assistance for self-management during daily life activities. Further development of the system should also include an automatic control of an infusion pump for a more accurate dosage or a delivery of an extra dose, when necessary.

The remote computing platform underneath REMPARK provides automatic data storage, processing, analysis and visualization tool so that other relevant actors, doctors and caregivers, can be included in the loop and can provide the best-informed, evidence-based and personalized action for therapy and healthcare management. In particular, REMPARK can be used to extract indicators of the disease evolution and open the possibility of a better adjust of the medication and treatments.

One of the main advantages of REMPARK is its alignment with the actual evolution of the care models, that are moving from reactive to proactive approach. The strength of the system relies on the ability to obtain, analyse and integrate data from different sources: sensors, patient reported data and healthcare professionals.

The system is able to receive the data transmitted from the sensors. For each individual patient, it is possible to define a range of normal values, permitting to stablish some levels of alarms according the thresholds defined by the professionals. Any deviation from normal event creates an alert on the system and consequently there is a reactive intervention. The DMS includes a shift management module that manages the tasks and users that are at each shift. The shift manager can see the workload and type of activities done in the centre for efficient management of the care process.

The REMPARK system is an interesting solution that includes a wide range of functionalities for delivering comprehensive and customizable integrated care by a multidisciplinary team, but it also enables the treating physician to maintain a supervision of the patient throughout the journey; from patient disease diagnosis, through treatment plan definition, customizing specific protocols. One of the main problem in most existing systems for disease management is that the patient's data used for treatment is collected on the system overtime and is not updated by other systems that are used for treating the patient. This problem is particularly relevant when the patients are affected by a multimorbidity.

Multimorbidity is commonly defined as the presence of two or more chronic medical conditions in an individual and it can present several challenges in care particularly with higher numbers of coexisting conditions and related polypharmacy. For this reason the patient is treated by the neurologist for the PD and in parallel he is treated by the general physician in the community for other health conditions. Typically, a completeintegration between data doesn't exist and there is a lack of information about the overall condition of the patient. Furthermore, the communication between the healthcare professionals treating the patient is quite limited. This way of delivering care results in a fragmentation, some duplication and lacks of coordination.

REMPARK, offers an integrated environment for data and knowledge sharing for all care providers and also it provides clinical guidelines and a decision support tool for healthcare providers helping in management of patient's disease. This tool enables treating several conditions at the same time in a coordinated way avoiding the interaction among drugs and the negative impact on the patients compliance.

Another additional value is the available integrated platform, allowing the stablishement of improved communication channels among patients and professionals (doctor, nurse, etc.)

Some advantages for the patients are the increase of their treatment performance scores, the improvement of the patients' empowerment and a good degree of satisfaction.

10.4 Limitations of the REMPARK System

The characteristics and advantages of REMPARK system would contribute, for sure, to the mitigation of the barriers analysed in the Section 10.2. It is obvious that some limitations exist, and because a main part of REMPARK is the sensor (a wearable device) for the detection and measurement of the motor symptoms related with PD, part of the inherent limitations are in line with those of different wearable products developed for their use in e-Health domains.

There is a need to improve the cost-to-benefit ratio of WT strategies and their effectiveness for chronic disease and rehabilitation therapy management. Innovative research is still needed to continue developing the best combination of wellness, special needs and technologies in order to assist and maintain the preferred QoL for individuals with chronic disabilities and older adults.

A further limitation is the need of a refinement of the technology to correctly measure and monitoring non-motor symptomatology of the disease. Patient priorities and sources of disability often arise from non-motor deficits, such as: depression, anxiety, fatigue, orthostatic hypotension, sleep disturbance. There is a need for developing unobtrusive systems to monitor non-motor end points in the home environments and community settings. As it has been presented along the chapters, REMPARK was designed as a complete system, where in a limited way, the consideration of the non-motor symptoms was already included and worked-on. Further versions of REMPARK system must, for sure, consider and emphasize these aspects.

Usability of the REMPARK system was a challenge from the beginning of the project and quite good results are reported in Chapter 9. A problem that the consortium was working-on and tried to solve is the PD patient's adapted smartphone interface. The special adapted interface (see Chapter 6) is a good point for the usability and acceptability of the system.

After the REMPARK project finalization, relevant experiences would be used for an improved version of the different parts of the system, but always some common issues related with the acceptability of wearables in general would be encountered.

Systems are often not as user friendly or compelling to adopt as they should be. Currently, patient and care giver engagement with wearable and mobile technology is modest, as shown by recent studies.

A lack of motivation to use wearables and monitoring systems should not be underestimated, particularly in the absence of meaningful feedback provided to their users. Preliminary evidence suggests that patient empowerment and their inclusion as active players in the development of research activities may favourably impact on compliance. As a preliminary conclusion, it could be stablished that additional research is needed to determine the characteristics and the feasibility of wearable systems for long-term monitoring of motor and non-motor symptoms that would be acceptable to patients.

10.5 Clinical Applicability of the REMPARK System

The power and usefulness of the devices and systems based on ICT technologies are still under-recognized by a part of the physicians and professionals participating in the treatment and management of the PD patients. Its potential, considered as an aid to the patients, is also under-exploited.

In the case of REMPARK, the system is based on non-obtrusive elements worn by the patient: the waist inertial sensor, a smartphone and an auditory cueing system. The elements seamlessly communicate with each other and with the server platforms providing further service and capabilities. REMPARK is an interoperable standalone solution that can be easily integrated in Hospital Management Systems and external eHealth services operated by third parties.

The REMPARK system core is a sensor device placed on the patient's waist that it is able, as it has been already explained, to identify and quantify the main motor symptoms of PD (bradykinesia, dyskinesia and FOG) in real time. This sensor also recognizes and registers the patient's motor fluctuations (ON and OFF motor states) in ambulatory conditions in a very reliable and automatic way while patients are performing their normal activities. Finally, the device wirelessly transmits this information to a server.

Furthermore, the use of these devices with the described characteristics would enable the doctors to accurately personalize medication intakes and, thus, improve the patient's response to the treatment. A system with the REMPARK characteristics would result an invaluable tool in the diagnosis and management of PD. A correct detection of the symptoms would help to, on one hand, enhance the effectiveness of the oral medication through a better regimen adjustment; on the other hand, to automatically control the administration of an extra dose when the patients are using infusion pumps (apomorphine or duodopa). These new approaches are aimed at significantly improve the QoL of patients and, it will allow a deeper understanding of the personalized evolution of the disease.

An additional benefit is represented by the contribution of the system to the real and effective implementation of a multidisciplinary care. In fact, clinical experience suggests that optimal management requires a multidisciplinary approach, with multifactorial health plans tailored to the needs of each individual patient. In case of PD patients, the multidisciplinary team includes physiotherapists, occupational and speech-language therapists, dieticians and social workers.

The use of systems highly based on ICT technology, like REMPARK, contributes to alleviate and facilitate the organization of the multidisciplinary approach improving communication between the different health professionals, the patients and the caregiver.

Another area that would benefit from a tool, like the sensor developed in REMPARK is the clinical and epidemiological research. This kind of studies are expensive and highly laborious. Sometimes, they suffer economic limitations that can affect the methodological rigor of the studies carried out. In general, the studies based on movement disorders are especially complicated, on the one hand by the lack of well-established markers to establish a clear diagnosis and, on the other hand, by the lack of uniformity in diagnostic criteria.

Finally, it must be pointed out that REMPARK system technological innovation can be an interesting challenge for companies developing business based on the commercialization of new e-health home based services for PD patients. Such tools would provide the unique opportunity to objectively monitor and control the treatment efficacy of any given therapy on an individual basis, which would sustainably change traditional health care methods.

10.6 As a Concluding Remark

We can conclude that REMPARK system is a big step to a new approach to a new approach of the PD treatment where the technology contribution is helping to provide a different and complimentary view of the symptoms of the disease by offering clinicians a complete map and evolution.

The provided information will not only be useful from a diagnosis but rather would contribute to a better and more effective management of the disease.

This kind of systems, when completely developed, will be relevant because the possible supervision and measurement of the evolution of the symptoms and the appearance of new ones. This will permit a rapid action when necessary, facilitating among others the establishment of preventive polices according patients' individual needs.

References

[1] Chang P. Wray L. Lin Y. Social Relationships, Leisure Activity, and Health in Older Adults, Health Psychology, American Psychological Association, 2014, Vol. 33, No. 6, 516–523.
[2] Wallace E. et al., Managing patients with multimorbidity in primary care. BMJ 2015;350:h176.

Index

219

About the Editors

Joan Cabestany holds a Telecommunications Engineer degree (1976) from the Universitat Politècnica de Catalunya (UPC) in Barcelona, Spain. He got his PhD (1982) from the same University, where he is working as a Full-time Professor in the Department of Electronic Engineering.

He has been involved in research and innovation activities as a part of his career. He is the responsible member of the AHA ("Advanced Hardware Architectures") research group at UPC, with expertise on reconfigurable hardware, electronic system design, advanced hardware architectures, microelectronics, and VLSI design. One of the main topics of interest has been the practical application of the Artificial Intelligence to the functional improvement of electronic systems.

He is a member of the CETpD Research Center staff structure at UPC since 2005. The activities of CETpD are focused on technological developments applied to the improvement and help people with chronic diseases, like Parkinson Disease and different conditions related with aging.

On behalf of UPC, Prof. Cabestany has been responsible for several EU-funded projects. Among them, the REMPARK project on PD management and the FATE project for the accurate detection of falls in aging people are the two projects, where Prof. Cabestany has been acting as coordinator. He is the co-author for more than 100 research papers and communications to Conferences. Prof. Cabestany is a co-founder of the Sense4Care Company (www.sense4care.com), a Spin-off of the UPC commercializing relevant innovative research results.

Angels Bayés Rusiñol MD is a Doctor in Medicine and Neurology specialist. Her career and research activity has been focused on the study and treatment of Movement Disorders like Parkinson's Disease (PD) and Tourette syndrome, and on dementias, like Alzheimer's disease.

She has been the director of the Unit for Movement Disorders and Parkinson of TEKNON Medical Center in Barcelona (Spain) during the last 20 years.

Her main research areas are related to the implementation of complete treatment, complementary therapies, psycho-education, and to try to improve the quality of life of patients suffering from movement disorders. During the last 5 years, she has been cooperating with an engineering team at Universitat Politècnica de Catalunya, in Barcelona, for the development and implementation of technological solutions for the support and management of PD.

She has participated in 39 research projects, with special mention of the EU funded EDUPARK and REMPARK ones. She has authored more than 60 publications, including 6 books, many conferences, communications, and training courses.